Architect's Notebook 건축가의 수첩

초판 발행	2013년 10월 14일
1판 5쇄	2019년 6월 25일
엮은이	담디
펴낸이	서경원
편집	이현지, 나진연
디자인	정준기
번역	이지은
펴낸곳	도서출판 담디
등록일	2002년 9월 16일
등록번호	제9-00102호
주소	서울시 강북구 삼각산로 79, 2층
전화	02-900-0652
팩스	02-900-0657
이메일	damdi_book@naver.com
홈페이지	www.damdi.co.kr

Fifth Edition Published	June 2019
Compiler	DAMDI Publishing Co.
Publisher	Kyongwon Suh
Editor	Hyunji Lee, Jinyoun Na
Designer	Junki Jeong
Translator	Eisa J. Lee
Publishing Office	DAMDI Publishing Co.
Address	2F, 79, Samgaksan-ro, Gangbuk-gu, Seoul, 01036, Korea
Tel	+82-2-900-0652
Fax	+82-2-900-0657
E-mail	damdi_book@naver.com
Homepage	www.damdi.co.kr

지은이와 출판사의 허락 없이 책 내용 및 사진, 드로잉 등의 무단 복제와 전재를 금합니다.

All rights are reserved. No part of this Publication may be reproduced, transmitted or stored in a retrieval system, photocopying, in any form or by any means, without permission in writing from DESIGNERS and DAMDI.

정가 22,000원

ⓒ 2013 DAMDI and DESIGNERS
Printed in Korea
ISBN 978-89-6801-022-4

Architect's 건축가의 수첩
Notebook
The Treasure house of Idea

CONTENTS

6	**Interview**
26	**Gutiérrez-delaFuente Arquitectos**
50	**CHA:COL**
70	**YOSHIHARA McKEE ARCHITECTS**
88	**Gambardellarchitetti**
98	**Donner Sorcinelli Architecture**
110	**ARHITEKTURA d.o.o.**
124	**Hybrid Space Lab**
132	**eu.k Architects**
148	**BOARD**
156	**b4 architects**
182	**Heather Woofter** [Axi:Ome]
194	**ARPHENOTYPE**
206	**OSA**
220	**AA & U**
232	**DIMOS MOYSIADIS and XARIS TSITSIKAS**
242	**EXTERNAL REFERENCE ARCHITECTS**
260	**OOIIO Architecture**
278	**Jongyeon Bahk** [Grid-A]
300	**object-e architecture**
316	**IaN+**
344	**LIMA URBAN LAB**
356	**ZO_LOFT ARCHITECTURE & DESIGN**
370	**IAD PARTNERS**
384	**Miguel Arraiz García** [Bipolaire Arquitectos / Pink Intruder]
406	**INDEX**

Interview

What is a Notebook to you?

What is a notebook to you?

A notebook is a memory stick. A notebook is a data base.
<Gutiérrez-delaFuente Arquitectos>

To me, a notebook is a natural extension of a mind that thinks and records visually. Technology, however advanced, still replicates at some level, the direct connection a sketchbook establishes between the hand and the eye. I think of it as an accurate record of our own growth and change. I often reach for my old sketchbooks to see how I approached a visual situation then and how I think now.
<CHA:COL>

SM: A notebook is full of memories, a journal, a place to jot down notes, reminiscences, ideas. HY: I have two types, one is a source book of new ideas, the other one is to study ideas; find and investigate solutions. <YOSHIHARA McKEE ARCHITECTS>

A sketchbook is a reserve. Represents a forbidden world where there are no rules. It's an accumulator from which to choose. It 's the place where things are manifested as if it were dawn.
<Gambardellarchitetti>

It's a way to focus my ideas, thoughts and visualizing schemes and concepts. More of them are plans, sections, perspectives, all together in a sort of bazaar of the architectural process. <Donner Sorcinelli Architecture>

Free drawing sketch is one of important tools of an architect, with which the architect explores the phenomena of a space and its various atmospheres. Unlike a painter, where the drawing/picture is the final result of his work, the architect's drawing functions more as his working tool and a method of discovery. One can only draw what one comprehends. Through observation one grasps the mechanics and psychology of a space and as one sketches his understanding of a space, it becomes a part of one's long-term memory and consciousness. This is the reason why one carries the sketch book at all times, and a mere look at its pages takes one back to the emotional and rational experience of the moment in time and the atmosphere in which the sketches were created. <ARHITEKTURA d.o.o.>

Drawings and analog notations are essential within the process of developing ideas, concepts and designs. As our projects are strongly process-oriented we often make notations of processes.
<Hybrid Space Lab>

Our notebooks represent our office (studio) itself. When a project begins, numerous sketches begin to fill up the wall. There are various kinds of sketches, from conceptual to furniture details, from presentation sequence to the architectural form; in between them are photos and study models. As we develop the project by discussing the ideas while walking through the spaces, the studio itself becomes a large 3-dimensional notebook. In other words, the project process itself is the organic notebook. Moreover, when we go out for meetings or lecture at schools, we use 'Satellite notebooks' which are actually regular notebooks. All the sketches and notes made on these notebooks are connected and supplemented to the 'Core notebook' which is our studio.
<eu.k Architects>

A notebook actually means very little to me. I am not using one. The only time I ever used a notebook was during my first year of studying architecture, when my teacher forced us to use

one. But I abandoned it very quickly, because I did not believe in its value. In fact, I considered it counterproductive and something that blocks transparency and the exchange of ideas during design processes within teams. In notebooks information becomes very secret, hidden and private. When I hear sentences such as "Architects use notebooks as their treasure chests of ideas", I think of architects of the 20th century such as Louis Kahn, who was known for making numerous trips to Europe during his career in the United States, sketching and writing down everything he saw in a sketch- or notebook, probably because he did not have pocket-sized cameras or smartphones yet. That is why notebooks appear to me today as something outdated and old-fashioned, something like a Friendship book in a time when everybody has a profile on Facebook, something like using a sketch as an architectural representation when we have renderings.
<BOARD>

A notebook is a safe-conduct for the mind and for ideas. The need is to take it everywhere, at any time you may need to use it to fix ideas, concepts, thoughts, visions. It 's still very versatile, more than any electronic tool available. And it is immediate, do not need of battery or energy to work, only your active brain. It should be light and easy to handle, ready to use. It can frame design thoughts or visions of travel. In the case of use in travel, it records visually what it is important for you, but through a filter that selects only what really matters, even just suggestions. Over time it becomes a repository of memory, also useful to recall ideas to reconsider even after a long time. Normally the paper is white, of a robust weight without signs or lines to allow maximum freedom of expression. The drawings or words are laid out with a pen usually thin and smooth, they are sometimes also used colors to emphasize certain concepts. The size is normally small and pocket, but often are also used common A4 sheets held together by a provisional clamp on a rigid support to use it everywhere.
<b4 architects>

My notebook is something to quickly add notes, hold ideas link comments from lectures and conferences to systematic ideas. A place were I note important books, people, phone numbers or even recipes. I have nowadays a notebook function in my smart phone, but it is not the same. Even so, that Arphenotype is at the moment very digital driven, I think the analog way is as important as it always was. The sketchbook enables me to react on the fly - often I am surprised about what I find in my books, when I start looking at them years later.
<ARPHENOTYPE>

Our work at OSA is digitally based, developing complex geometry or computational systems of variable relationships. Therefore, the use of a sketching and the notebook has changed simultaneously as the design process has changed from 20 years ago when we were starting out in the design field. This is not to say that the notebook is any less important, but rather that the information it provides to initiate or develop a project has changed. Today, I find it more difficult to sketch out an entire building idea in plan, section or perspective since the quality of space due to the geometric complexity is difficult to establish in a drawing. If the design is less complicated, such as a chair with less geometry, I can still use the sketch to attempt to work out the main design ideas. But, typically I use sketches to diagram relationships between spaces or systems. There are two basic types of sketches I use as a design tool; analytical and relational. Analytical sketches either interpret existing conditions or project more specific qualities about a space or a series of objects while relational sketches attempt to map out variables in a system to speculate on the range of possible associations between components.
<OSA>

My notebook is my mind. By the time something is imprinted in the notebook has the possibility to be remembered and to go through a process of transcription into design. The process of imprinting an idea in the notebook becomes an integral part of the design process. Everything is noted in order to allow a return to it, a recollection or a reconnection to other current thoughts. Such thoughts may take the form of a drawing, or the form of a text. Such thoughts may have a

documentary character, a sort of mapping of actual conditions which I find either in physical environment, in books or in presentations of architects' work. In other cases, they may have a projective character, meaning they become an imprint of possibilities projected in the future. That could relate to initial thoughts about projects, such as the sketches attached relating to an architectural competition for the re-conversion of a former agricultural farm into an incubator of activities.
<AA & U>

XT: It is a companion that allows you to be productive any time. The ideas are not only products of long hour focusing on a designing task. The incubation of an idea can awaken our enthusiasm on our day off or during lunch break. A sketchbook and a pencil are essential at that time.
DM: Sketchbook partly carries the notion of identity. There is a magical power that make everything drawn in a sketchbook more important than any of the delicate strokes on plain paper sheet.
<DIMOS MOYSIADIS XARIS TSITSIKAS>

Undoubtedly, the notebook to the architect is the meeting point of human and handmade aspects within a project. Within it, thinking and production develops in real time without technical intermediaries.
But, delving into what the pages of a notebook can offer, we think of it as blurred thinking territory, which surrounds the intellectual integrity of the architect. In the sketchbook not only do you find construction details or delicate perspectives but also thoughts, quotations, bibliographic references and an endless insight into the designers' life. Designers' sketches paradoxically merge with their daily lives (dates to remember, lists of materials for a model etc.). In this way, the end result when a book is completed, is close to a so-called 'artist's' book which makes it a medium of infinite possibilities: a time of freedom; being able to fill your pages with the representation of the passage of a professional life-time, the exceptional chance to go back in thought processes and ultimately, the possibility of reading spatiotemporal visual discourses of architecture.
In short, the architect's notebook is the object in which the thought process is captured with abstraction, where it is at its most fully contingent, both in absolute size and in its micro manifestations expressed through rhythm and cadence. <EXTERNAL REFERENCE ARCHITECTS>

I do not have a notebook; I collect papers, models, ideas, etc. in boxes. For me those boxes keep the project soul.
My notebook is any white piece of paper or board that I can write, draw, cut, etc. I do not have a physical notebook, my notebook is a mountain of papers that are always all around me… included little sketch models that I produce constantly (I do not like work with the computer on 3D). I prefer to model with my hands. It gives always a "human touch" to every project.
This big collection of random documents is my notebook, which I collect like a little treasure, because it shows all the project process: first ideas, shape and program distribution evolution, rejected options,… and also short sentences that I wrote while thinking on the building showing he worries I have during the design process, that always influence on the final result, like "let´s push harder", "I need to call to this guy", "I need to pay this bill", "do not have time", "at 18:00 h I will meet that guy"…etc
A notebook is a piece of your live, and it is made of lots of different things, that we keep in boxes.
<OOIIO Architecture>

It's a thought process. I record every moment of my thoughts in my notebook. During the school years, it was a book full of lecture notes. These days, it sometimes become presentation material to clients and team members when discussing about the projects throughout the design process. At times, it became my drawing portfolio, full of sketches and my thoughts and reminiscence on the memorable architectures that I visited. During travels, it also became my

diary with admission tickets and receipts taped onto the pages.
<Jongyeon Bahk [Grid-A]>

I use notebooks usually in order to record thoughts and ideas the time that they appear. For me, most of the times, it works better with text than with drawings. In that sense it is always good to have a notebook with me so whenever something comes to mind I can write it down. Usually I forget about what I wrote quite fast, but then after some time I go back to the notebooks, look through them and find again those ideas. In most cases they just stay there, in the notebook, but some of the thing might actually find their way into one of my projects or my texts.
<object-e architecture>

My notebook is a fundamental instrument in my daily work. It helps me materialize ideas, concepts and thoughts.
<LIMA URBAN LAB>

Once Walter Benjamin said "Do not miss any thought, and keep your notebook as the authorities keep a register of foreigners". Well Actually it's like that! A sketchbook it is like a partner or a collegue that helps you reminding your ideas, or to preserve an emotion before it becomes too "reasoned". But it's also like meeting an old friend after a long time. Looking through it you will remember a lot of things you had probably forgot about your life.
<ZO_LOFT ARCHITECTURE & DESIGN>

If my notebook were computerized, it would be an external hard drive, a memory asset of my thoughts concerning projects and diverse architectural situations. But it's a paper pad that must be a "Moleskine". Regardless of the size or thickness of it, I have an exclusive relation with this reference. If I happen to forget it, a piece of paper tablecloth, a bag or a napkin will do. Each of these drawings will end in my Moleskine notebook anyway.
<IAD PARTNERS>

Is my personal diary, the place where I put my ideas during the time. Sometimes it becomes a book, because I read it again after some time and find old ideas that were not developed at that time, but ideas that now can be used and re-developed. <Miguel Arraiz García [Bipolaire Arquitectos / Pink Intruder]>

Any episodes or memories related to a notebook?

When you listen to a song, sometimes you can trip back in the time and remember perfectly a situation or a period of your life when you listened it first time. Having a look into the old notebooks, you can have similar feelings, you can also trip back on the time, and remember perfectly the moment where you did the sketches or the architectural way you were exploring at that moment. It's a time-machine.
<Gutiérrez-delaFuente Arquitectos>

I keep separate sketchbooks, separated by size and paper type for each time I travel. It keeps future organization precise and simple so I can draw in the moment, wherever I feel like on each sketchbook. The separation of sketchbooks helps me connect very directly with memory and

association. I often add a small word specific to that place or time, so I can recall why I drew a particular detail or chose an angle.
<CHA:COL>

SM: When traveling, people would peak over my shoulder and ask what I was doing, I could share or quietly close the cover. While working in Japan, it as a great way to communicate, if you drew it you were sure that the other person would understand, words and talk were to easy to get lost in translation.
<YOSHIHARA McKEE ARCHITECTS>

When I began to study architecture gave me fifty notebooks green pocket. I thought that I would have been enough for a long career. Ten years later they were all filled with the things seen, to imagine things, of things which I then built and there was no free space on a white paper.
<Gambardellarchitetti>

Episodes are more related when I forget it, so I have to use papers of cafeterias, post it, coasters and everything can be ready to be drawn. These makeshifts becomes attached parts of the notebook, thanks to the pocket at the end of it.
<Donner Sorcinelli Architecture>

Three years ago we participated in the international competition for four new boulevards in Benghazi, Libya. We only learned of the competition three days prior to its conclusion. As we stopped for dinner on the way back to Slovenia, my colleagues tried persuading me that there is no point in entering the competition due to the lack of time. In addition, we did not have our computers with us. It was then that I picked up a thick marker and drew the concept of a solution on a tablecloth. Three days later we won the competition. <ARHITEKTURA d.o.o.>

We use communal notation sessions - within the office but also in workshops with external participants. These communal notations are tools to merge analytical investigations with synthetic design processes and to create a networked mental space of interwoven concepts and ideas.
<Hybrid Space Lab>

We sketch ideas related to the projects at home as well. Those usually end up on the wall beside my desk as well. So ultimately, the wall at my house becomes a 'satellite notebook' as well. One day, my seven year old son was staring at one of the sketches on the wall and said, "I am going to draw that." Then, he grabbed a trace paper and started to trace over the sketch. After a few months, he looked at a model, drew it on a piece of paper, and pasted a tree made out of real wood. He has made a hybrid expression by combining a 2D sketch made from 3D model and a 3D object together in one. Even he is sketching in organic methods.
<eu.k Architects>

As I am never using a notebook and have rarely used one in the past, I cannot divulge any episodes or memories related to it here. Nevertheless, I use sketching but not in a notebook, but mostly on a sketch roll. That makes the sketches generally rather temporary and of an ephemeral nature, as they are usually thrown away directly after they fulfilled their purpose, which is mainly to get a step further in a design, research, or study process. It does not matter if they are rather ugly, as long as they communicate the right thing. But they don't have to be representational anymore, meaning that they are not much used for final presentations, as they were in the past, before the time of CAD, Adobe, and rendering software. Sketching can become much more enjoyable than it was in the past as a lot of pressure has been removed from it. Today, it becomes ever more clear that an important effect the integration of the computer has had on the architectural design process has less to do with form, organic shapes and complex geometries, as once was hoped, but with the liberation of the sketch from the notebook, where it was damned to

permanence and burdened to represent.
<BOARD>

Over the years they have been preserved with chronological order all the notebooks in various dimensions. It happened that one of them has been lost in the classrooms of the Faculty of Architecture in Rome a few years ago. It contained all the sketches of a journey between Berlin and Canada, with important experiences recorded in the drawings and in the notes. The sense of to be out of element for the loss was great and not yet fully ridden out: part of the 'archive' of the mind is lost. The feelings, the ideas recorded on the architecture visited in Berlin (Mies, Libeskind, Eisenmann, Scharoun, Piano, etc..) and Canada have been and still are an important cultural baggage for the project activity. From that event periodically all the sketches are scanned and stored digitally, to avoid further unpleasant similar events.
<b4 architects>

Each notebook represents a different time. In reviewing drawings of specific places, I am often transported into the past and I remember the intangible qualities of a place in addition to its form.
<Heather Woofter [Axi:Ome]>

It may sound strange, but I have the feeling that the best ideas are born, when I do not have access to the internet - maybe we are all slaves of this digital cloud already. At the moment I am commuting every week between Cologne and Berlin - which is a 4 1/2 hours train ride, one way. Here in Germany we do not have internet yet on the trains. So it is like being free, time to focus on reading books and working. Here is my notebook central element of my mobile office setup.
<ARPHENOTYPE>

I was moving my family from the east coast of the United States to the west coast (Los Angeles) and we drove across the country, stopping in Utah to see some amazing rock formations. I created a series of analytical sketches to interpret and project a sense of scale to the fairly ambiguously scaled, but massive objects. In addition, I had brought a child's toy camera similar to a Polaroid camera that could create instant photos, but they were only 3 x 4cm. I pasted the photos of the rock formations in the sketchbook to establish a dialog between the photographic recording and my analytical interpretation. I was not interested in the accuracy of how the rocks existed since the photo served that purpose. I was more interested in the qualities of scale, mass, and hierarchy that I could project onto them through the process of sketching. <OSA>

I have realized that my notebook is my mind, when I went to a lecture by a fellow architect and I had forgotten to take it with me. During the whole presentation I was imagining how I was imprinting ideas in my notebook, starting off a process of transcription. My sketchbook has become a valuable interface between myself and the world allowing the projection of imaginaries and their link to any kind of precedence.
<AA & U>

DM: A particular booklet is the one that I have in my mind as my first sketchbook. It actually was. After the childish drawing blocks I was given by an architect, teacher of mine, a sketchbook and I was told to carry it everywhere. This was enough to be convinced. A strong bond is created between the one that holds the pencil and the sketchbook. I remember myself filling the pages with tremendous speed literally turning pages every minute. It was like a contest with no opponent. Today the pages are filled after deep thinking. Still fast but each of the strokes is meaningful. Even draft sketches of an architect reflects something more than the ideas represented.
<DIMOS MOYSIADIS XARIS TSITSIKAS>

Interestingly, the architect's notebook, read after some time, always produces a certain tenderness in the sense that within its pages, the designer relates to their surroundings with more

freedom, open-minded thinking, fearlessness and innocence. In this way, it becomes a kind of dialogue of censored options that have never been subjected to materiality, or the monstrosities designed without passion, that end up discarding the concrete reality of construction, economy and program.
Therefore our notebooks give off a certain rebelliousness and we discover the testimony of how through dreaming, in every moment, the spirit of each project is generated. I have come to realise over time, one can reach a maturity akin to the original sketch, although in many cases surprisingly far in form and materiality, it remains substantially tangent to its intellectual genetics. No doubt, if we were to tell the story of the office, we would dust off each sketchbook, and find individual reflection places, representing individual concerns within collective Professional consensus. And at this point, there are exceptional events as the drawings demonstrate a time where the hand of a member of the practice is at the service of the thoughts of another. A true professional emotional time.
<EXTERNAL REFERENCE ARCHITECTS>

We use any kind of paper to draw in, and usually advertising papers that we recycle, or wasted printouts from other projects, etc.
So at the end of the day, when you see the box-sketchbook for a particular project, you discover that we have used the test printouts for that other project that we were designing at the same time on our office, and suddenly lots of remembers of that time come to your mind, people that was working on the team, old project options, etc. and the most incredible think is that you discover how the projects done at the same time on the office have some design connections, that you didn't realize at that time. But now you can see watching those old documents.
<OOIIO Architecture>

The most important aspect that I consider when purchasing a bag or going on a trip is whether my notebook can be easily carried and taken out for use. When I recently visited Niagara Falls, I was worried about my notebook getting wet, more than my camera or myself.
<Jongyeon Bahk [Grid-A]>

I remember at the beginning of my practice that I used to make my own notebook. I bought some white sheets and cardboard.
<LIMA URBAN LAB>

Probably too many. We keep all the notebooks we had in our life, since the first one. So each one eof them can be used to talk about us while growing up.
<ZO_LOFT ARCHITECTURE & DESIGN>

I have sometimes thought on delivering a whole project in sketch format. I think it would be such a time saving as the essence of these projects can be found in my notebooks. At the end of a meeting in some far away country, I like to seat in a bench (my public office in some ways) to take over with drawings the different points of view of the meeting. Without Wi-Fi or scanner, I just send some pictures of my notebook to the studio as a first debriefing. Long live Sketchbooking!
<IAD PARTNERS>

All my sketchbooks have a memory, they're always a present that someone has made me, and so they're really personal belongings.
<Miguel Arraiz García [Bipolaire Arquitectos / Pink Intruder]>

When and where do you use your notebook the most?

We use the notebook the most when we are out of the office, in our trips... but also at home.
<Gutiérrez-delaFuente Arquitectos>

For work and travel. At work, I keep a small notebook next to my desktop for quick doodles and sketches. For travel, I have a set of notebooks separated by size and flexibility, depending on the size of my backpack.
<CHA:COL>

SM: In class to "talk visually" with the students, as a travel diary of our trips and at work to develop an idea.
HY: For studying; when I need to find a specific solution, mostly at the office for a project we are working on, but anywhere As a source book for generating new ideas that have not been seen before; when I am very relaxed and inspired by something. Sometimes they are random thoughts. Mostly at home but anywhere, any time.
<YOSHIHARA McKEE ARCHITECTS>

Now I use a lot of my notebooks during traveling or when I do boring meetings at the university.
<Gambardellarchitetti>

Anytime and anywhere.
<Donner Sorcinelli Architecture>

I use the notebook mainly as a means of communication with clients. With the help of a drawing I can immediately make every thought tangible, one that has previously only been a part of my or client's perceptions, wishes, intentions. I perceive my hand as a tool though which the subconscious translates into conscious, the intangible into tangible. Such a sketch is also easily understandable to a client. This is why, at the end of our session, we often reach a mutual agreement, a visible result, to which the client agrees according to the principle: what you see is what you buy.
<ARHITEKTURA d.o.o.>

Special for our office are the communal creative tools of blackboards. In our studio we use large-scale monumental blackboards to develop ideas together. These are communal creative - sketching, noting and mapping - tools. As we draw and wipe and redraw and reinterpret ... 'new readings' and misinterpretations open up a creative space of unforeseen synapses, nurturing new ideas and innovative understandings. While working together on our blackboards, the most creative moments are the moments of accidental overlaying where new ideas pop-up and evolve.
<Hybrid Space Lab>

The most times I use a so called notebook is during architecture studio critiques at schools. In order to give feedback based on the students' materials, I need to remember their project process and my comments for every critique I have with them. So, I note everything down inside the notebook.
However, since we always sketch with various design materials at the office, especially together at a large conference table, the large sketchbook is always short on pages.
<eu.k Architects>

Ever since the sketch was liberated from the straightjacket of the sketchbook or the architectural notebook, it could be produced everywhere at any time. If I am on a plane or train, the corners of a newspaper are, for example, great places to write or sketch something that you wish to remember. However, I do most sketching in my office during the day. This activity is mostly driven, apart from the need to communicate ideas to others, by the fear of losing thoughts and ideas. But that kind of sketch shares the same destiny as the one on a sketch roll: once it has fulfilled its purpose and was translated into a digital architectural drawing, diagram, or image, or communicated to another person, it is thrown away. Nevertheless, I don't think we should mourn the short-lived nature of contemporary sketches; just as we do not mourn the short lives of mayflies that only live from a few minutes to a few days, depending on the species. Adult male mayflies have two penises and during the few days they live in spring or autumn they are everywhere, dancing around each other and copulating in large groups on every available surface.
<BOARD>

The beginning of each new project normally involves a phase of brainstorming that occurs to subsequent sketches on paper until you come to a concept that is checked as soon as possible to the PC. Subsequently, the sketches are a tool that continues to take place alongside digital design. The places where the notebook is used are always different. It happens that the most informal places are more fruitful for design ideas, particularly in the more intuitive phase: a café bar, the stairs in a public place, park, traveling on a train or even on holiday. The office or more 'official' places of the work are used for the process in depth analysis of ideas.
<b4 architects>

I keep a notebook with me throughout the day to make notes and think through design strategies. Equally, the notebook is a place to visually remember a moment.
<Heather Woofter [Axi:Ome]>

I use the notebook whenever it is suitable to hold ideas or input - specially at conferences, lectures or when I am invited as Guest Crit. At the beach, when I am reading a book to make references, to note quotes. When I am travelling, such as on planes, busses or in the train. Sometimes the best ideas appear in the weirdest situations, like waiting for the night bus after clubbing.
<ARPHENOTYPE>

I mostly sketch at my desk, though it is not always in my sketchbook. Therefore, I use my sketchbook half of the time at my desk and the other half when I travel since its portability makes it easy to carry.
<OSA>

My notebooks have become over the last years very light so I can carry them in my backpack wherever I go. Usually I have more than one at any given time and they are organized around anything that I do, with the front cover being a sort of an index of such activities. During the last years, both drawing and writing, have become equally important in filling the pages. Quite often they are scanned in the digital environment becoming the base for design processes or writings about architecture and urban design.
<AA & U>

It could be even in the car, if not while driving. It is a corny statement but you keep notes at any time since you woke up and you have not slept yet. Though where we both use our sketchbooks the most is in the offices while working.
<DIMOS MOYSIADIS XARIS TSITSIKAS>

Perhaps to concretise the notebook in a specific time and place is the opposite of its genuine

definition. One of the hallmarks of professional and personal identity of an architect is to be always accompanied by his notebook. Thus, the almost existential need of an architect drawing can not be separated from the notebook. Thus, the place and time to use the drawing does not exist. Any time is right to reflect drawing. Even while sleeping; how many times does a designer wake up in the middle of the night to draw sketches, often incomprehensible, generated in their night dreamlike drifts. Sometimes even when one sketched on a napkin in a bar or on a wall with a piece of chalk, sketches become precarious. Undoubtedly, the architects notebook is full of these napkins, nurturing an even greater vitality.
<EXTERNAL REFERENCE ARCHITECTS>

I try to use the sketchbook only at my studio or when I am working outside (client presentation, construction site, when I am on a lecture or exhibition, etc) I try to NEVER think on work when I am on holidays or with friends, or so. Notebooks are forbidden then!
<OOIIO Architecture>

For the past couple of years, I've been sketching interesting spaces and architectures as I travel around a city or a country. I also use a sketchbook for every project that I work on and record all the process in it. So, I tend to use it all the time, since I have a sketchbook with me when I travel or work at the office.
<Jongyeon Bahk [Grid-A]>

Most of the times, I use notebooks when I don't have access to a computer. Long trips, boring lectures etc. Also when I need to escape from the computer and think and draw for a little while in a different way. But in general, you can never know when you are going to need a notebook, so it is good to have always one with you.
<object-e architecture>

The best place to use a notehbook is simply everywhere! You can never know when you would like to point down something or when you'll need to show an idea to someone. The best thing is to have different sized sketchbooks to always have one in you pocket!
<ZO_LOFT ARCHITECTURE & DESIGN>

Out of the office, where the size of the notebook and his practical size make perfect sense.
<IAD PARTNERS>

Especially at home and during the night. I'm not good having only a sketchbook, I have several around the house. Sometimes it's messy because when you try to find an old idea you have to find the pieces of this idea in different sketchbooks. But it makes it more fun when you revisit your sketchbook.
<Miguel Arraiz García [Bipolaire Arquitectos / Pink Intruder]>

What influence does a notebook have in your projects and life as an architect?

We think that more than the notebooks, what it's really important in or architectural practice are the sketches. The notebooks are the perfect tool to drop in the same place many cross-ideas, sketches, references, memories...
<Gutiérrez-delaFuente Arquitectos>

While we both pride ourselves in extensive use of digital tools, I find it hard to imagine our creative process without a physical sketchbook or even a roll of trace around. Sketching by hand is as much a mental process as it is a physical process. I believe the process of sketching goes beyond recording. It sharpens your thinking as you draw.
<CHA:COL>

SM: Provides an overview, a time line, I still have my sketchbooks form University and travel through them to see how my preoccupations and views have changed.
HY: The problem solving sketchbook is a tool that directly affects our current projects, the source book works future projects and competitions.
<YOSHIHARA McKEE ARCHITECTS>

Drawing is important because it is a storehouse of things that the human brain can not remember. From these things, the project is purified. In a project I do there are many tracks in less than those that design.
<Gambardellarchitetti>

It's the way to maintain the route with coherence, driving the process and sediment ideas for the future.
<Donner Sorcinelli Architecture>

I was once considered to be good at drawing. I was accomplished in the renaissance style of pencil drawing. I used to love travelling in Italy and sketching old towns and palaces. However, this was mannerist drawing, an end in itself. With it one can only draw the architecture of a bygone era. An architecture replete with plastic ornament, reliefs, the play of light and shadow. Later on I noticed that this kind of drawing or sketching only removes me from the creation of contemporary architecture so I rejected it. Today I only use sketching as a working tool (and also to search for an atmosphere).
<ARHITEKTURA d.o.o.>

We sketch and make notations within the whole process of project-development. The blackboards are especially helpful as they help develop unexpected synapses, open up minds, unlock ideas, foster creativity.
<Hybrid Space Lab>

A sketch is an important factor in communicating ideas and coming up with a solution. A project called 'Gu-san-seo-ga' (library village in Gusan town) was a collaboration project with another office. During the process, many sketches were used to communicate with each other. In some cases, the partners of each office came up with one sketch together.
Sometimes a physical notebook can touch or simulate someone. A notebook filled with the process of the students' work leaves a deep impression to the students.
<eu.k Architects>

However ephemeral and "ugly", the sketch itself has still a very strong influence on my projects and therefore also on my life, and I believe the same applies to a lot of other architects of my generation; by contrast, the sketchbook has very little influence. I think it is no coincidence that the moment in the mid-1980s when computer-aided design programs appeared, the last notebook manufacturer, supposedly one of the original "Moleskine" producers in France, stopped production. It took a lot of years before an Italian company brought that kind of sketchbook back on the market, establishing it as a trademark, breathing new life into a dead product, and marketing it as something legendary that famous avant-garde artists and writers used in the past. Recent impressive growth numbers have proven its commercial success, which, I believe, is the reason I am now writing about sketchbooks to begin with. I think the sketchbook is still

essentially dead, but enjoys a boom in sales, because of brilliant marketing and branding to a nostalgic mass of people that wishes to look as creative as the avant-garde that was supposed to have used it in the past. Perhaps this reflects a certain desire for something permanent and solid in an increasingly ephemeral temporary digital world in which one click can destroy everything within a split second.
<BOARD>

In the evolution of the office's work, sketch have registered an important role for the development of the project. In life as an architect it becomes the privileged way to observe and communicate, a tool that becomes an inseparable part of themselves, whether it be a way to observe and record external things or inner visions. It becomes almost an extension of the body and for this reason one can not do without and you bring it everywhere in daily life. It 'a tool to get to know spaces and materialize ideas.
<b4 architects>

The notebook is a place to think. I believe there is a connection between the hand and spatial thinking.
<Heather Woofter [Axi:Ome]>

The sketch book is the 2nd step within formulating a system - first step is always the brain. Sometimes an idea is born and rethought over weeks, before I start sketching. Then it even can rest for years in my sketchbook before I work further on it. Once it is imbedded in a 3D modelling program the sketch process is evaluated on tracing paper, working models etc. - not really in the sketchbook itself anymore - unless it is a theoretical project.
<ARPHENOTYPE>

Even though our design process focuses on digital methods, I find sketching to be an essential design tool from establishing initial ideas through the project development and into construction. Sketching is essential as a way for me to measure my own design decisions as well as a means to communicate ideas quickly and visually to others. My notebook in particular, is valuable for me to evaluate and measure initial design ideas that are too young to be fixed where the real value of the sketch relies on the inexact qualities that suggest a range of possible interpretations. It is within these potential interpretations that creative ideas cultivate.
<OSA>

During the last ten years my involvement with architecture has been both through drawing and writing. The sketchbook as I mentioned already, becomes a sort of register of thoughts during the different stages of projects which could have a drawing or a writing form. I always carry a backpack with me so I could put inside my notebook.
<AA & U>

XT: From my point of view the sketchbook has no influence in the way that the design proceeds. It might only affect the speed. Also there no importance if the cover is leather or dressed with fabric since the paper sheets work well with the pencil that you carry.
<DIMOS MOYSIADIS XARIS TSITSIKAS>

Using a notebook everyday as a professional tool, inevitably offers a constant critical attitude to our projects that never inhabit the world of digital processes.
Moreover, the notebook offers the project a delicacy of reflection in terms of materiality and construction processes, as freehand drawing is a process of physical and intellectual construction itself. With all this, a sketchbook gives the architect the ability to stay in touch with the traditional tools of the profession, which capture the most humanistic aspects of the project.
<EXTERNAL REFERENCE ARCHITECTS>

Our box-sketchbooks are the real project. For me the project is not the final result, this is just a moment of the project. Actually the building is what happens since the first client meeting and the today's day user that is getting in or using it... and the box keeps the soul of how the building was born.
<OOIIO Architecture>

Although one piece of paper is very thin, these pieces come together to make on book, and as numbers of these sketchbooks piled up, they became a large volume. Sketching and recording thoughts about architecture became a habit. As time passed by, I started to think to myself and define what beauty is and what kind of space is a good space. I think it is very important to know, as an architect, what is beautiful and how to design spaces. The habit of using sketchbooks helped me define myself as an architect who knows.
<Jongyeon Bahk [Grid-A]>

I am not really fond of that romanticized idea of the notebook, that it is the place where the architect creates his projects, how his ideas are conceived etc. It is the same concept of the paper napkin and the sketches you make on them. I don't believe that projects come out of sketches made in 1 minute on your notebook or any other piece of paper that you have available. My projects are conceived and developed through drawings, models, 3d models, scripts and simulations that most of the time happen on the computer and are the result of intense work, and a process that is developed over many iterations. The sketchbook is a tool that complements all the above. It is useful in order to record ideas in relation to the projects and processes that you are working on, but it is never the main instrument for them. It might be that their use is more psychological than really functional.
<object-e architecture>

An architect communicates through graphic expression. This is why the notebook helps us to always exercise our ideas with diagrams, mental maps, drawings, etc.
<LIMA URBAN LAB>

Notebook in Arabic language means "well-ordered", so it should be like a step or a part of designing, and we were probably expecting to do that while using it. The fact is that our notebook will never be that tidy!
<ZO_LOFT ARCHITECTURE & DESIGN>

My notebook doesn't have an influence on me, but I do have an influence on it... It's sometimes a power relationship between us... Especially when I end one or start a new one.
<IAD PARTNERS>

It gives me freedom to think. When you start wit other process in the design (computer, calculations,etc..) everything seems to be controlled not only by you, but by the technical process.
<Miguel Arraiz García [Bipolaire Arquitectos / Pink Intruder]>

Are there anything else other than a notebook that you use to keep a record of your thoughts and ideas?

From last time, we realized that with the new technologies (ipad, smartphone...) we are supplementing (most of the times replacing) the use of the notebook with the use of these devices, these digital notebooks. You can make notes, write thoughts, but also you can take pictures, save internet references, interchange images... tons of images that are part of your memories when you are going to think in a project or any other architectural concern.
<Gutiérrez-delaFuente Arquitectos>

We both use digital tools extensively. Mostly for experimentation, organization and post-processing. We have a pen and tablet for digital drawing and a host of software -- one for recording written notes, one for digital doodles and one for organizing scans and images. I experiment with a combination of 3d modeling, vector drawing and raster work.
<CHA:COL>

SM: I some times use a lined notebook, but it is way too restricting, and I have tried gridded paper but it was also too restrictive. I unfortunately am
very messy and sometimes end up with things upside down, so if I stick to a sketchbook, although I have tried many shapes and sizes.
HY: For my source book I sketch on recycled paper used on one side. I can be more relaxed than on new white paper which seems so precious. I can draw anything sand do it poorly. Actually these sketches almost always turn out to be the best.
I do not need a "book". I just need paper and a good binding system to store it. For studying after some reflection I typically incorporate the ideas into the real drawing then I throw my sketches into garbage.
Sometimes I do want to preserve my first impression and come back to it while the project is developing and changing, so those I keep as scraps of paper, drawn on whatever.
<YOSHIHARA McKEE ARCHITECTS>

When I'm in the studio I love to draw and compose on large sheets of drawings invention completing or reassembling images taken from the internet. Just because the design is related to the things I've seen and I like to see right through him.
<Gambardellarchitetti>

Everything can be drawn or marked, also the sand on the beach.
<Donner Sorcinelli Architecture>

It is my opinion that, in the course of his work, an architect should use all the possibilities offered by various tools. I combine sketching with computer drawing, architectural models, photographs, video and sounds. All of the above tools are continuously present on my desk and in my work flow I leap from one form of expression to the next.
<ARHITEKTURA d.o.o.>

Other than the traditional notebook, I recorded my thoughts in various mediums, such as books, writings, blog, and webzines. Recently, I began to use twitter, facebook, and other SNS mediums to leave thoughts and communicate with others regarding them. All these move 'together.'
<eu.k Architects>

In addition to the abovementioned corners of a newspaper and sketch rolls: used sheets of paper and paper from printing errors are also great media to sketch and write on. I do that a lot.

In my office we once had, for example, an A3 printer and a lot of A3-sized paper. But ever since the printer broke down, the sheets of A3 paper remain unused in the storage, piling up without being used. I recently counted around twenty boxes of five hundred A3 sheets of paper each. So what we do is, we cut those sheets into handy A4s and use them as sketch paper, and to keep a record of thoughts and ideas as well. But then again, those papers are usually thrown away after the sketches and thoughts are translated onto another medium. Only a handful are kept in a project folder. All this shows how in my office the architectural sketch is demystified and used merely as a tool to remember and communicate things – something utterly temporary that does not require prestigious and glamorous treatment.
<BOARD>

Also the camera is a powerful tool to record feelings and visions store. And it is immediate, versatile and it records a substantial amount of useful information. But photography in the hands of an architect has less 'selective' chance of the topics that interest in confront of the possibilities of the drawing, and a most limiting expressive possibility for the ideas to communicate. Of course it is a very personal and expedient opinion of course.
<b4 architects>

Photographs help to record, but the act of drawing creates a kinesthetic memory, and for me, has a greater possibility of emotive qualities.
<Heather Woofter [Axi:Ome]>

Yes, they are all over the place. I have to admit, I am not an organized person with my ideas, sometimes I use books for sketches, napkins, physical models and so on - sometimes the original ideas is taken up at another location, being aware that the systematic idea stays the same.
<ARPHENOTYPE>

I sketch quite often, though it may not seem like I do because I don't always use my notebook. I sketch a lot on post-it notes and random pieces of loose paper. Really, I sketch on whatever is closest to me when I have a thought. The problem is since those sketches are not bound in a book, they tend to get lost after a while or when I move onto the next project. They should be kept and pasted into a notebook since it is a record of your thought process.
<OSA>

There are two other ways of keeping records of thoughts and ideas. The first is a specific about design process and the second one is generic regarding references of all sorts. The former is about using physical models as records of ideas about specific projects. Series of models from the very beginning of the design process materialize design ideas. Quite often it is one model that thanks to its easy manipulation is materialized out of a continuous overlaying of gestures during the various stages of the design processes. The second way of keeping records is through digital archives of images from travels in various cities. They are well registered so I could have easy access to them when I feel there is something to find during a design process.
<AA & U>

We both have spend tons of A4 format plane paper sheets. One sided sketches or two sided right before the recycle. It is been sometime that we keep our digital tablet sketchbooks. Not as handy but still it brings communication a step forward. Our working together for many years now make the fancy sketches unnecessary for the conception of an idea. The traditional sketchbook is more official in a way, though personal, the means to convince oneself rather than a client, a statement throughout time.
<DIMOS MOYSIADIS XARIS TSITSIKAS>

In addition to the nostalgic enrichment of the notebook, the immediacy of communication that today's technology offers is invaluable. On one hand, through smartphone technology, communication between members of the office have almost surpassed the format of e-mail communication to impose ´chat´ applications. Thus, the smartphone allows us to share images and scenarios when practice members are not together.

Our iPhone has almost become a sort of 'collective eye' or 'common memory'. When we're apart and when there is a reference that interests us, architectural or not, or when work problems arise, a photograph sent by chat and a brief dialogue which is often only understandable by ourselves, become an essential professional communicative mechanism. So, the message history, photographs and thoughts in our messaging thread, is a way of consolidating a kind of common digital sketchbook that we can access by sliding our finger across the touchscreen, just as we turn, the yellowish pages of our old notebooks.
<EXTERNAL REFERENCE ARCHITECTS>

As I said, we do not have traditional notebooks on our studio. We collect every kind of notes: at one side we stack papers and papers with ideas, printouts, data, etc. on the other side 3D sketches made randomly with cheap materials for our volumetric studies, on the wall in front of us we hang the selected ideas and they become like "project design guides" o "design directions".... And at the end, when we finalize our work we collect everything on boxes, that we keep as little treasures.
(always in every document that we write, we put the date and the project, for the final collection).
<OOIIO Architecture>

Although I use my smartphone sometimes, with all the various digital tools these days, I still like to write and draw with my hands. In between the pages of my sketchbooks, you can find odd pieces taped into them; they are papers from when I forgot to carry a sketchbook with me.
<Jongyeon Bahk [Grid-A]>

As I already said, I usually use digital computers to keep track of my ideas and thoughts, especially when they have to be expressed through drawings and images. When it comes to texts however, I use my notebook maybe to the same extend with my computer.
<object-e architecture>

I use my sketchbook most of the time.
<LIMA URBAN LAB>

Everything you can write on. For real. The problem is that sometimes you need to copy them because you can not take the table or the wall with you!
<ZO_LOFT ARCHITECTURE & DESIGN>

No, I have an exclusive relation with my notebooks, almost obsessive. If I don't have it close I feel the same way everyone else feels without their smartphones... I am sure you get what I mean.
<IAD PARTNERS>

I always try to sketch at the same time in 2D and 3D, so I usually work with conceptual models. I keep those ideas in 3D around me during the whole process of the project. Apart from that, once the idea of the project is in an advanced process I love to contact an artist to make his own interpretation of the work. Those objects created became a part of my daily life.
<Miguel Arraiz García [Bipolaire Arquitectos / Pink Intruder]>

Architect's

Gutiérrez-delaFuente Arquitectos

www.gutierrez-delafuente.com

The Gutiérrez-delaFuente Arquitectos office was founded in 2006 in Madrid by Natalia Gutiérrez Sánchez (Madrid, 1980) and Julio de la Fuente Martínez (Madrid, 1980).
The founders are graduate architects of Escuela Técnica Superior de Arquitectura de Madrid (ETSAM) and have continued their training in Madrid and Ateliers Jean Nouvel office in Paris. Natalia Gutiérrez was a Member of the Ethics Committee of COAM (2010-2012).
Currently Natalia combines the office with her job as town planner in a town of Madrid.
They have won prizes in numerous national and international competitions, including two in Europan 09, Europan 10, Europan 11, IQ Wohnquartiere and the Multipurpose Centre Valle de Salazar.

They have obtained several awards, highlighting the Bauwelt Award 2013. They have been published in numerous magazines, exhibited in Spain, Germany and Austria as well as having several models exhibited in The Architectural Gallery of the magazine El Croquis.
They have been guest professors at universities in Sweden and Germany.
Gutiérrez-delaFuente focuses its research activity in contemporary urban mutation processes such as the reconversion of abandoned brownfields and industrial areas, or the phenomenon of shrinking cities. Also, they are especially interested in the new ways of living and housing developments.
At present they are building several singular projects in Spain and Germany, and also working in Austria.

What is a notebook to you?

당신에게 수첩이란 무엇인가?

A notebook is a memory stick. A notebook is a data base.

수첩은 메모리스틱이다. 수첩은 데이터베이스다.

Any episodes or memories related to a notebook?

When you listen to a song, sometimes you can trip back in the time and remember perfectly a situation or a period of your life when you listened it first time. Having a look into the old notebooks, you can have similar feelings, you can also trip back on the time, and **remember perfectly the moment where you did the sketches or the architectural way you were exploring at that moment.** It's a time-machine.

수첩에 관련된 에피소드가 있다면 들려달라.

어떤 노래를 듣다보면, 그 노래를 처음 들었던 순간을 정확하게 기억할 때가 있다. 예전 수첩들을 볼 때, 비슷한 기분을 느낄 수 있다. **처음 그 스케치를 그렸던 순간, 그 때 추구하였던 건축을 기억하게 된다.** 수첩은 타임머신이다.

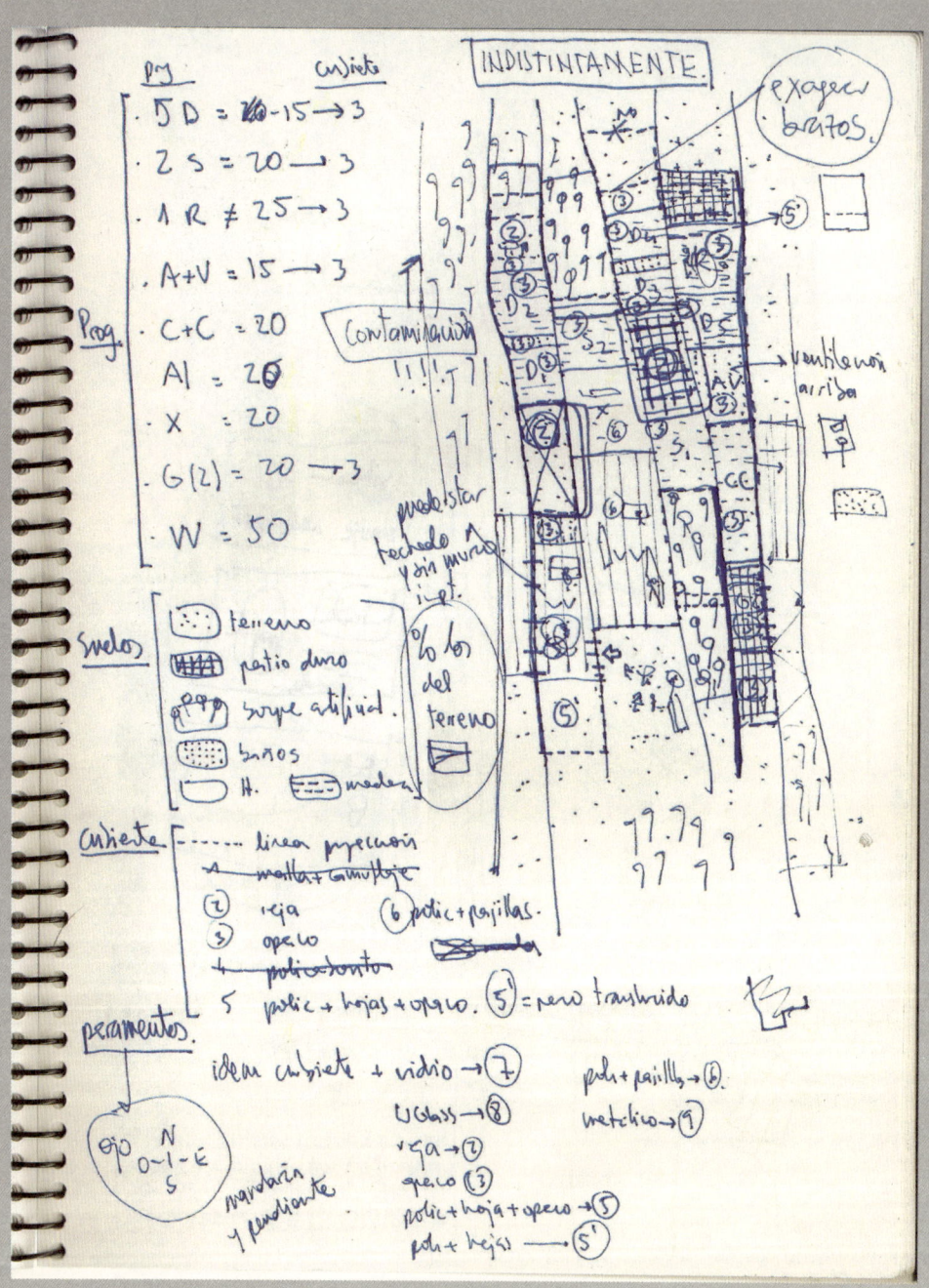

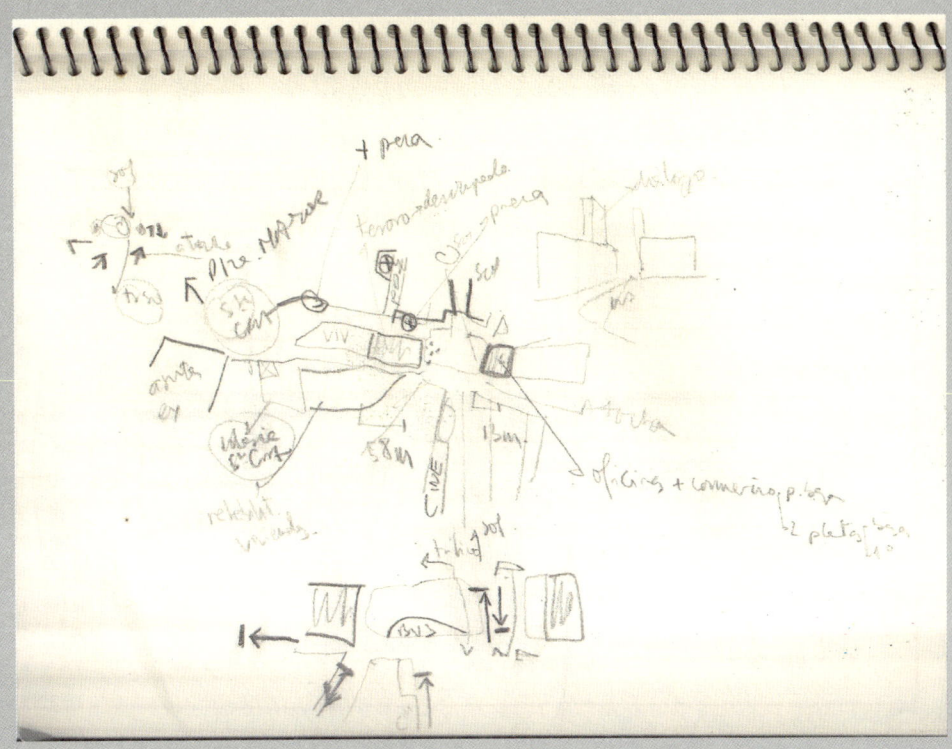

When and where do you use your notebook the most?

We use the notebook the most when we are out of the office, in our trips... but also at home.

수첩을 가장 많이 사용하는 공간과 때는?

우리는 여행을 갈 때처럼 사무실 밖에 있을 때 가장 많이 사용한다. 가끔 집에서 사용하기도 한다.

You can remember perfectly the moment where you did the sketches or the architectural way you were exploring at that moment.

What influence does a notebook have in your projects and life as an architect?

수첩은 당신의 작품과 삶에 어떤 영향을 끼치는가?

We think that more than the notebooks, what it's really important in or architectural practice are the sketches. The notebooks are the perfect tool to drop in the same place many cross-ideas, sketches, references, memories...

건축에서 정말 중요한 것은 수첩 그 자체가 아니라 그 안에 들어있는 스케치들이라고 생각한다. 수첩은 다양한 아이디어들과 스케치, 참고 자료들, 그리고 추억들을 한곳에 모아두기에 가장 좋은 도구다.

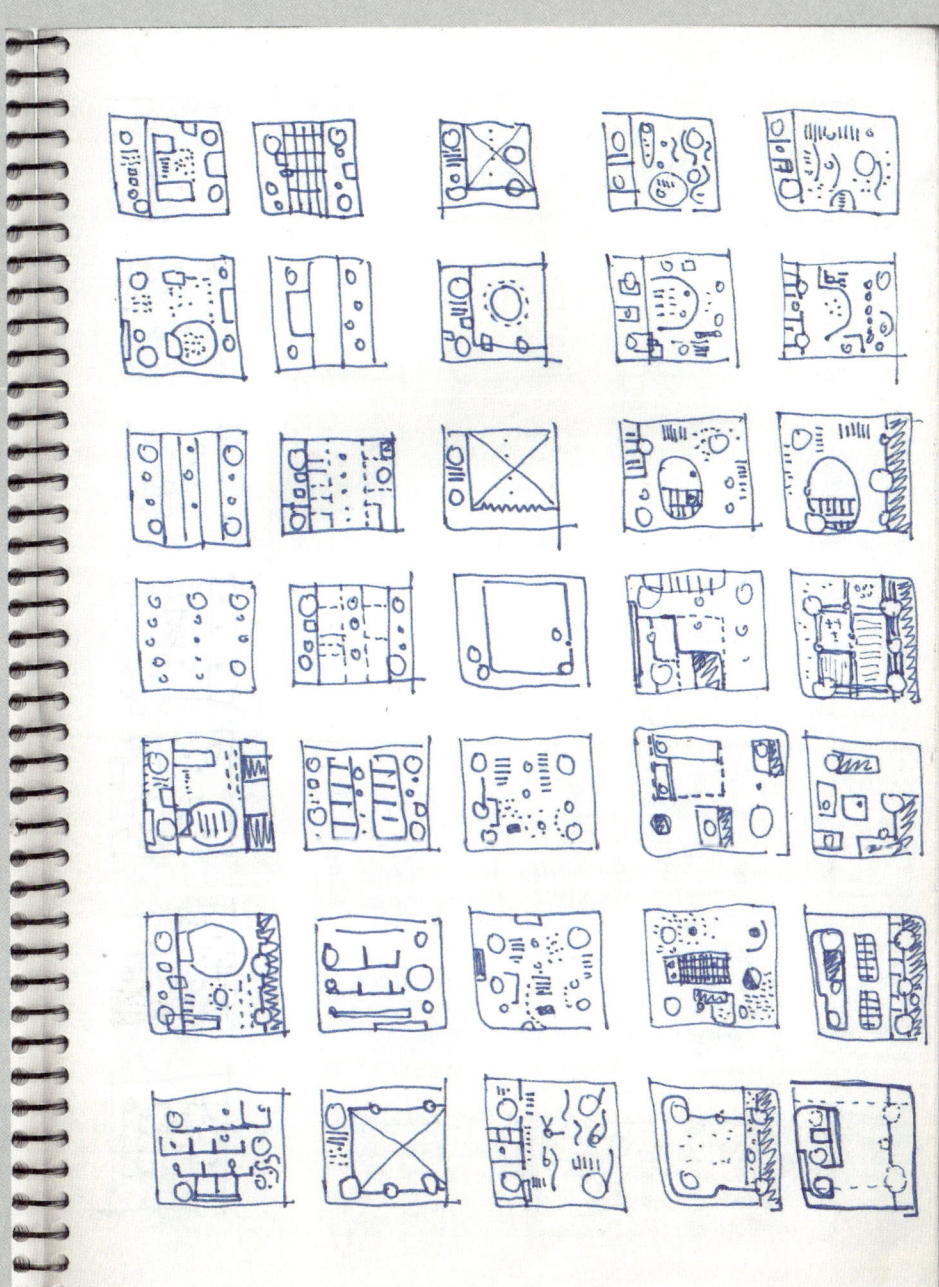

Are there anything else other than a notebook that you use to keep a record of your thoughts and ideas?

수첩 외에 자신의 생각을 기록하는 방법과 도구는 무엇이 있는가?

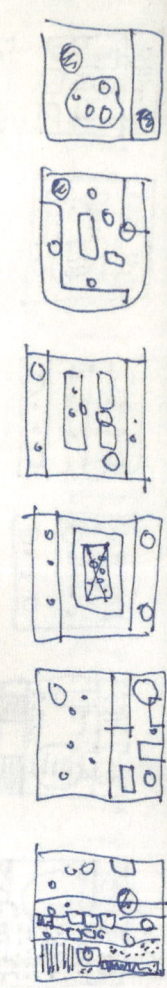

From last time, we realized that with the new technologies (ipad, smartphone...) we are supplementing (most of the times replacing) the use of the notebook with the use of these devices, these digital notebooks. You can make notes, write thoughts, but also you can take pictures, save internet references, interchange images... tons of images that are part of your memories when you are going to think in a project or any other architectural concern.

요즘은 새로운 기술들이 (아이패드나 스마트폰) 우리의 수첩을 보충 (대부분은 교체)하고 있다는 것을 깨달았다. 이러한 새로운 기계들은 노트를 하고 생각을 적을 수 있을 뿐만 아니라 사진도 찍고 인터넷 자료를 저장하고 이미지들을 바꿀 수 있다. 프로젝트나 다른 건축 문제들에 대해 생각할 때 기억 속에 남는 많은 이미지를 저장할 수 있다.

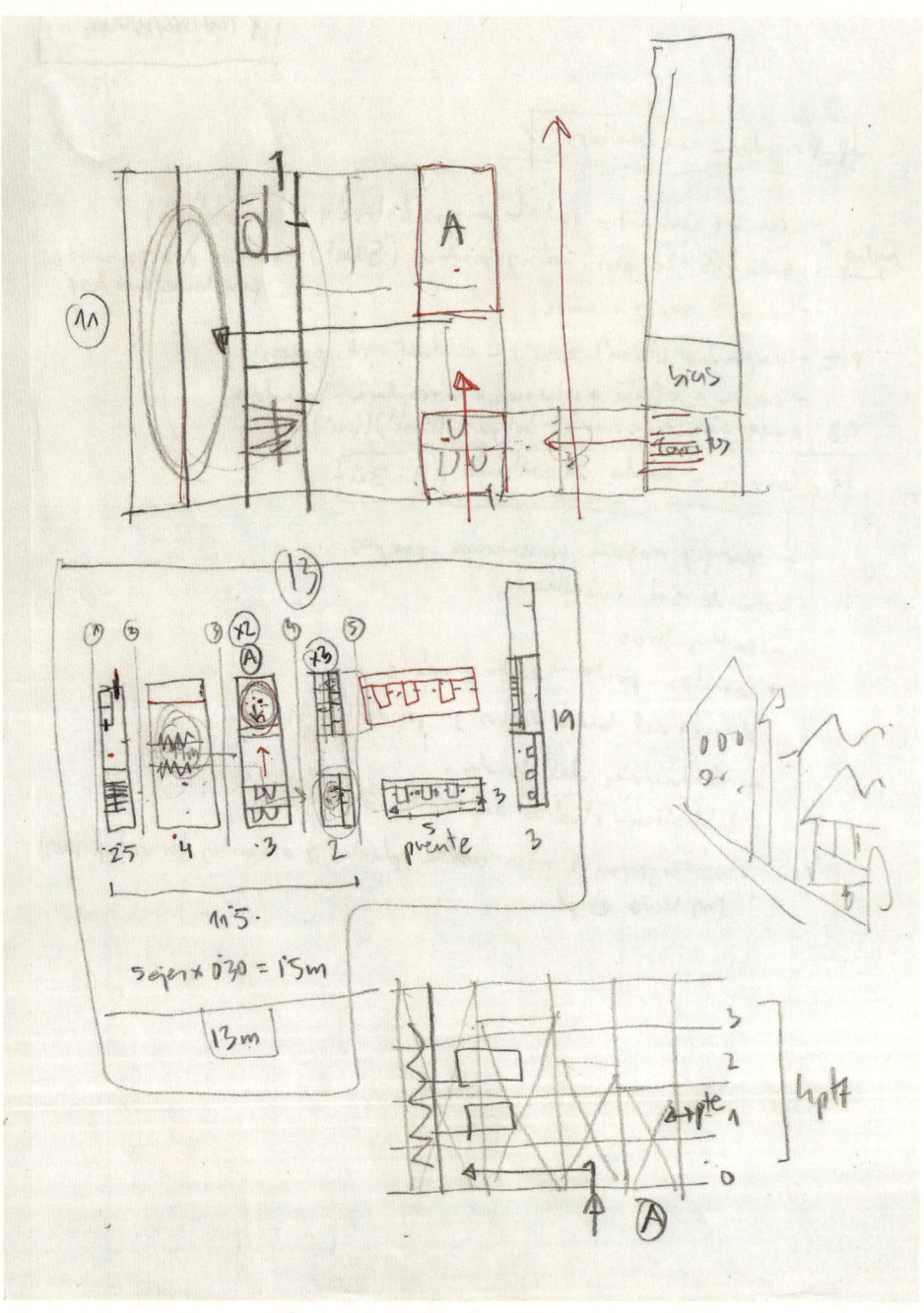

***Childminders Centre**, Selb, Germany*
Authors: Gutierrez-delaFuente Arquitectos + TallerDE2 Arquitectos.
Local office: SelbWERK GmbH.
Photographer: Fernando Alda.

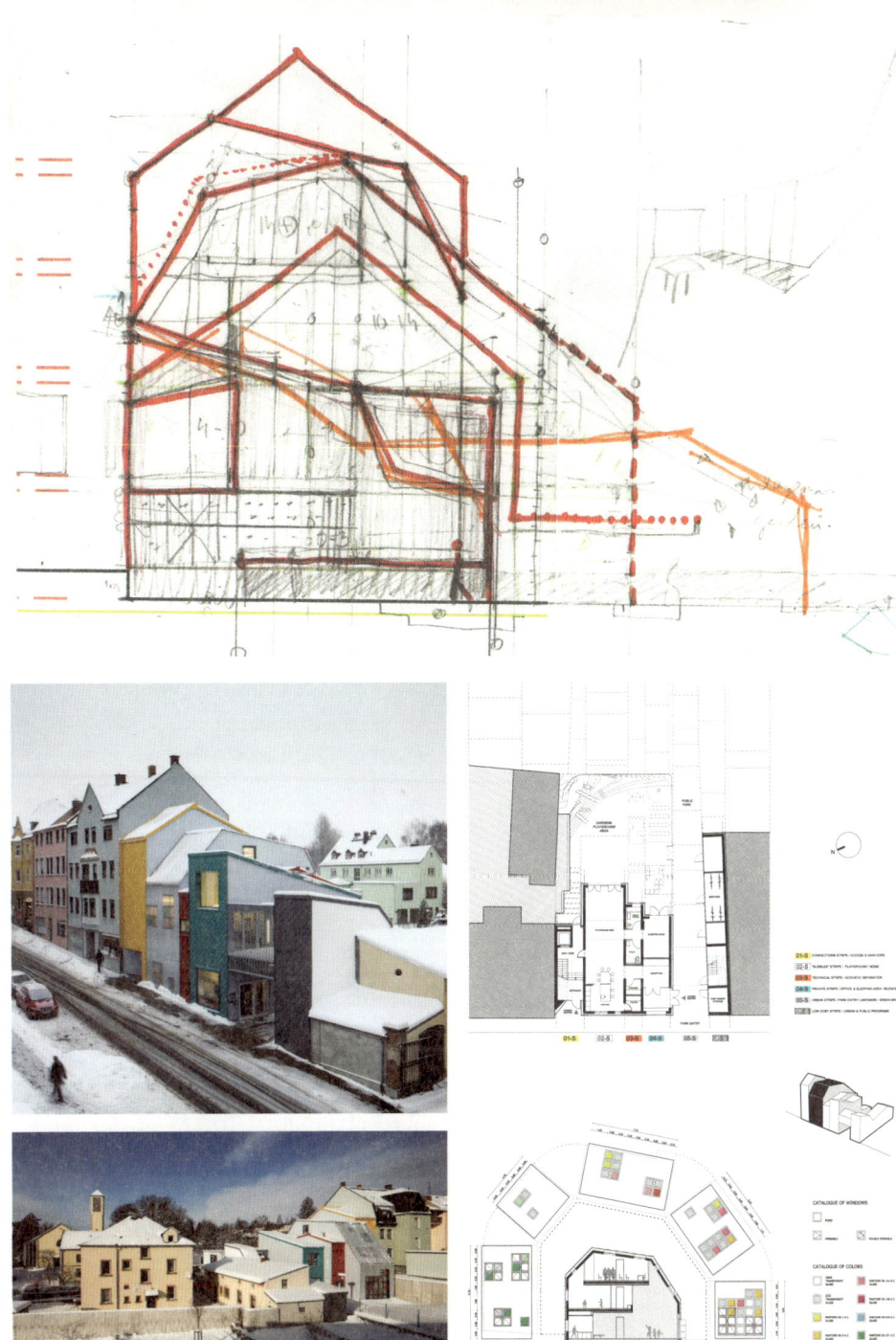

***Multipurpose Centre**, Bornos, Spain*

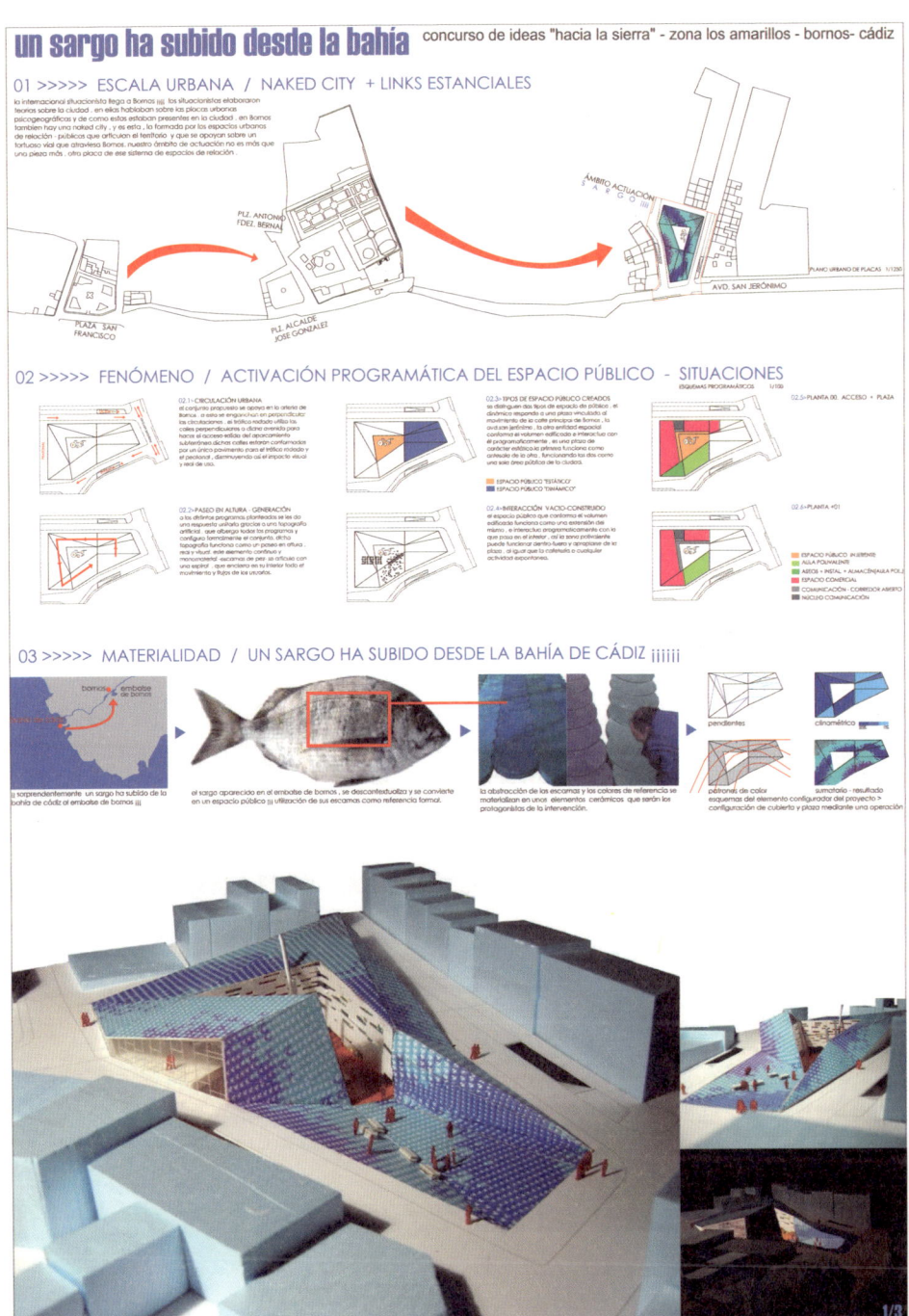

We think that more than the notebooks, what it's really important in or architectura practice are the sketches.

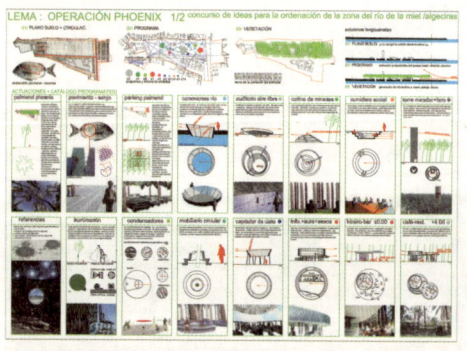

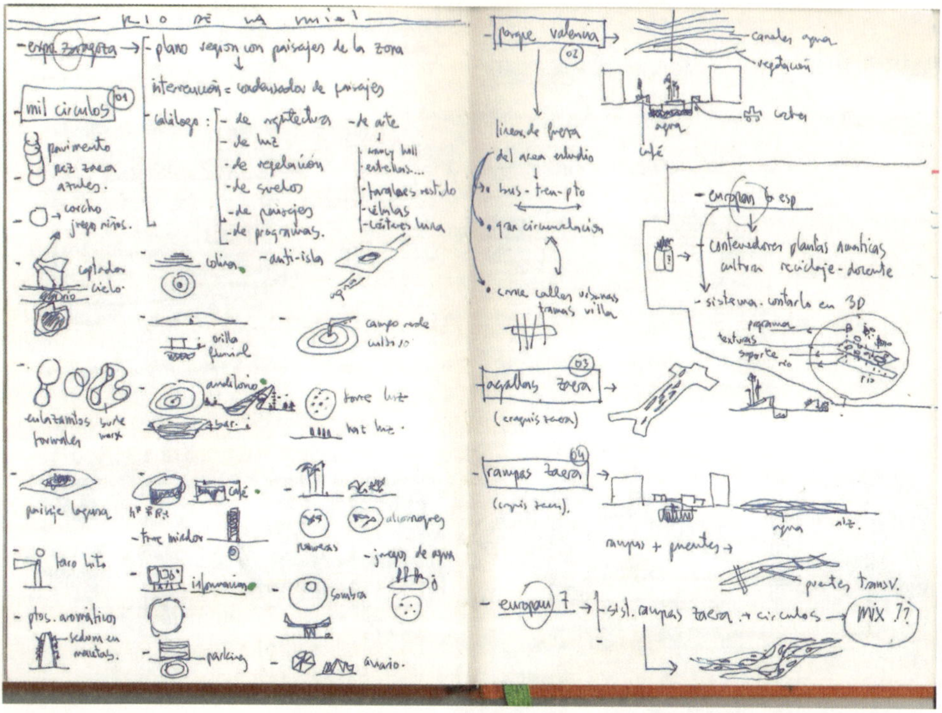

*Planning and urban design of *Rio de la Miel*, Algeciras, Spain

***Urban acupuncture**, the innercity of Madrid, Spain*

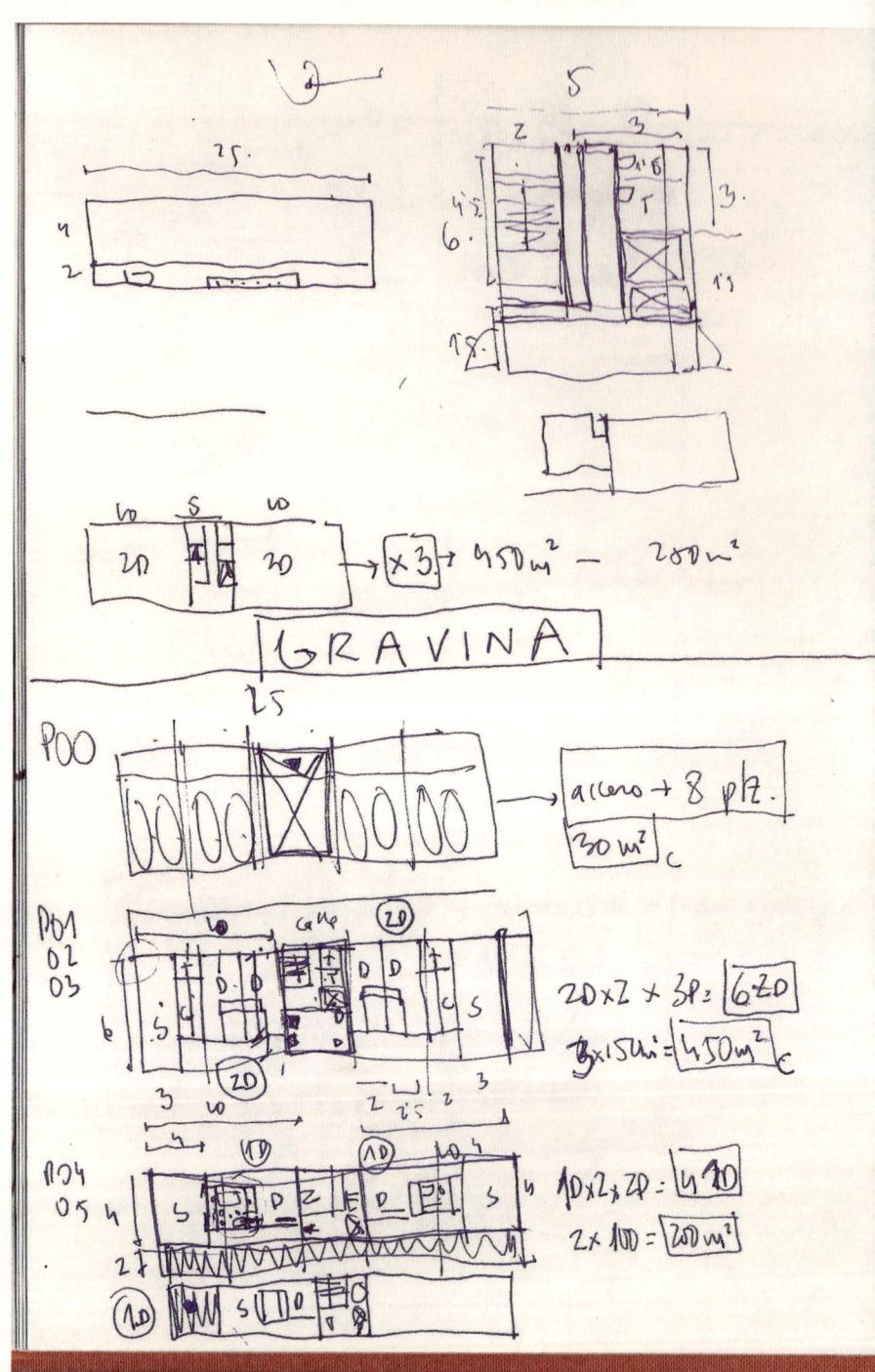

Gravina, Madrid, Spain

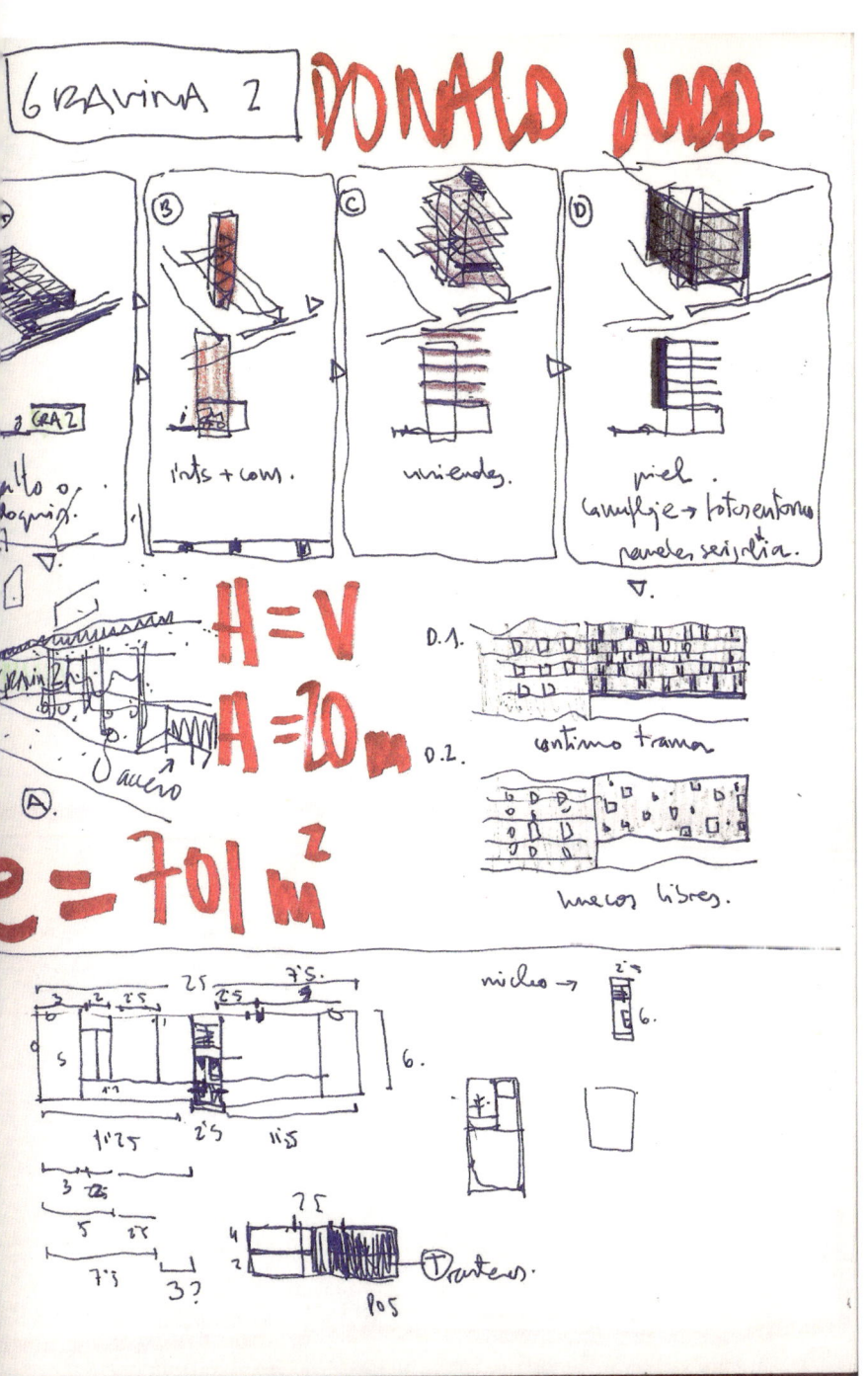

Plaza Fuencarral, Madrid, Spain

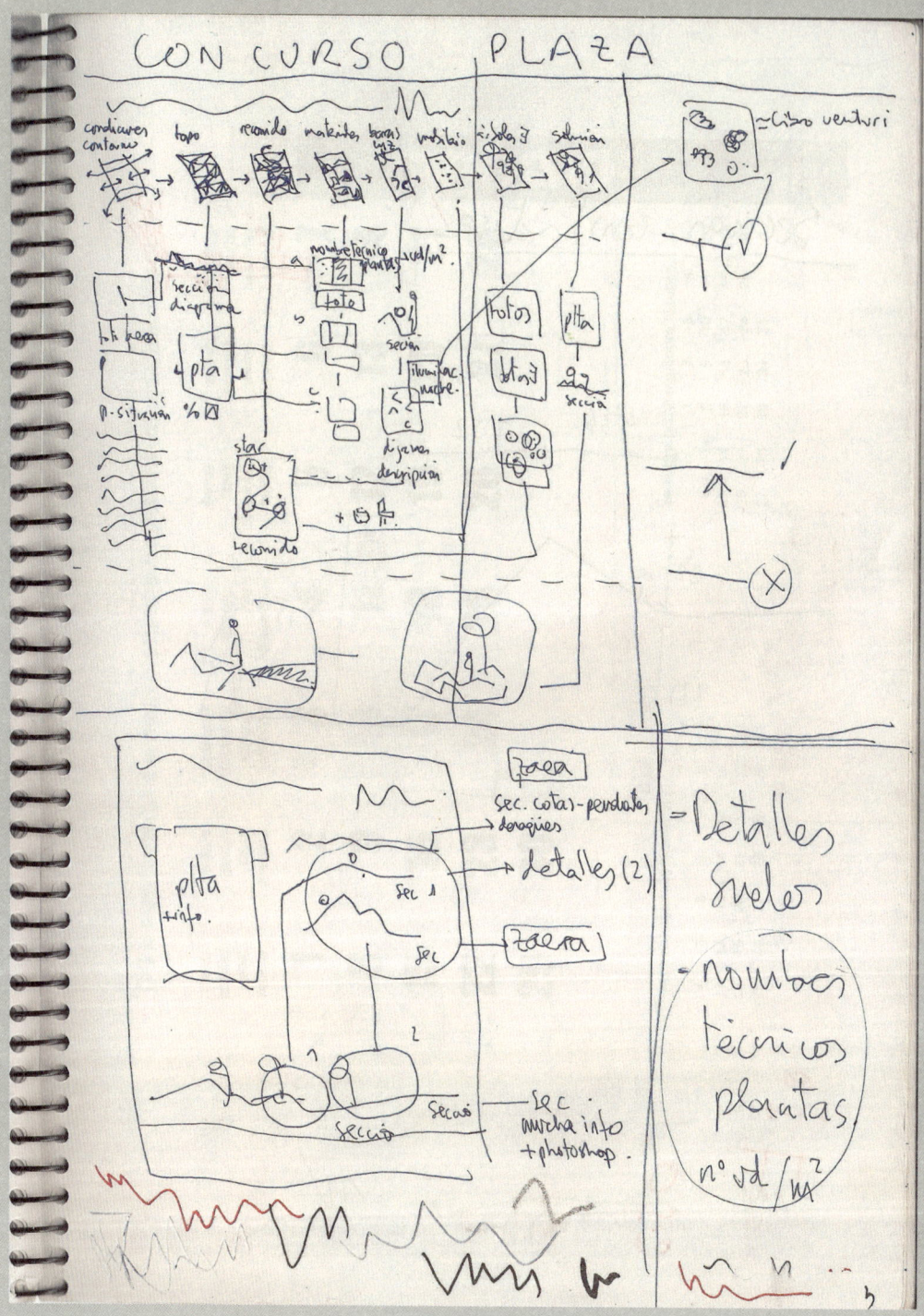

University Building, UAM, Madrid, Spain

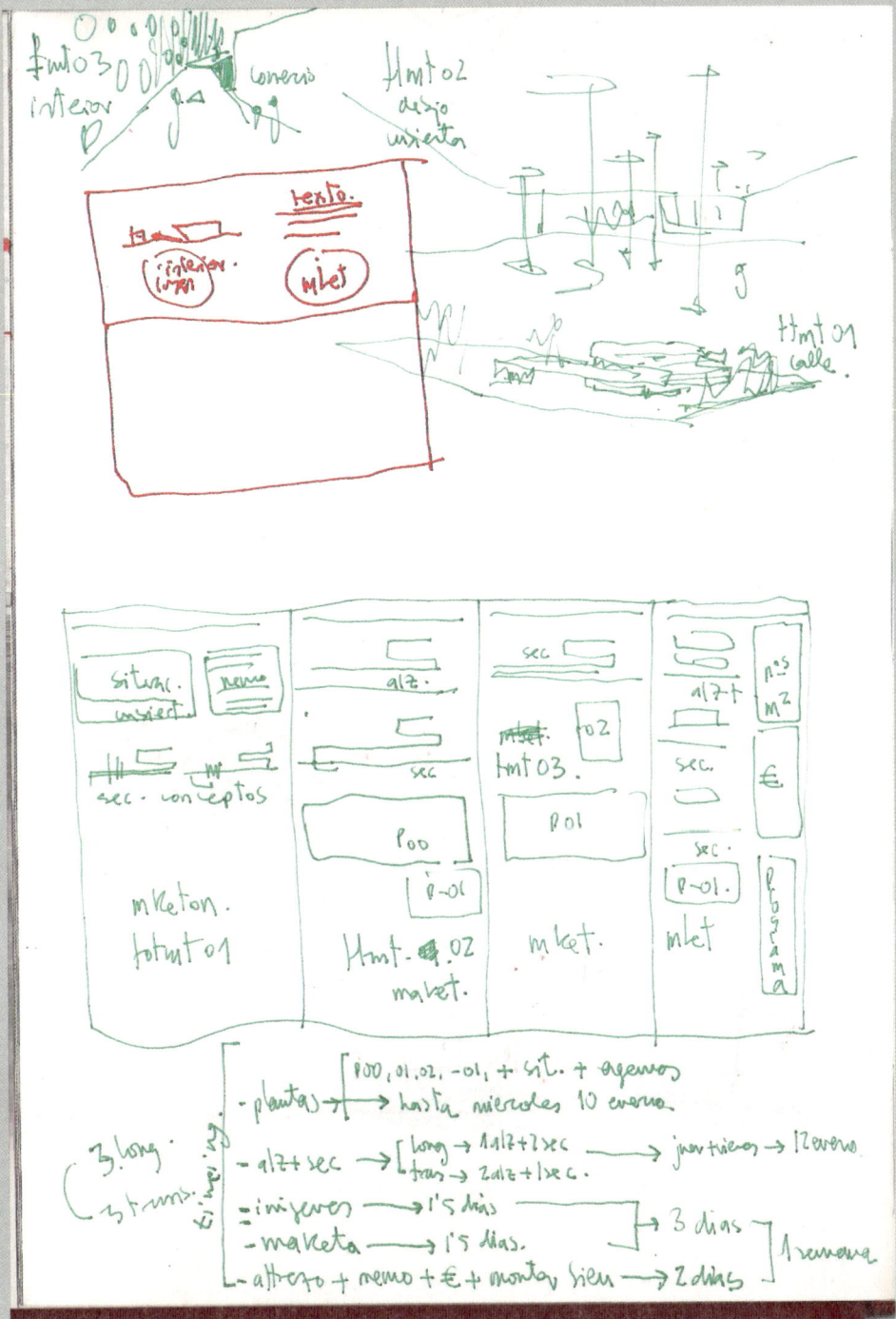

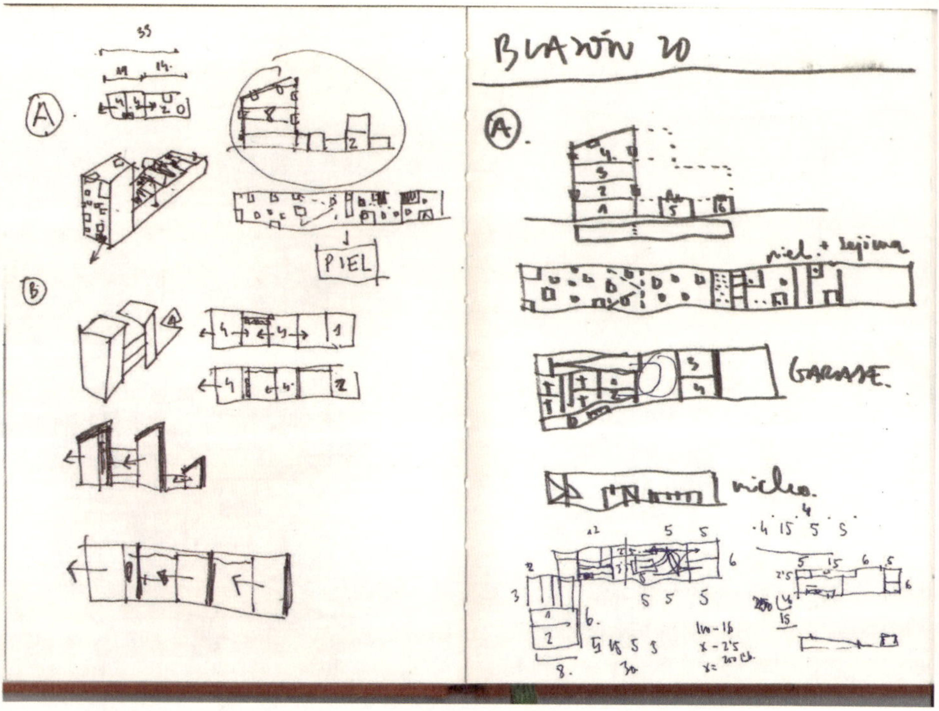

Carabanchel, Madrid, Spain

Vallecas 15.1, Madrid, Spain

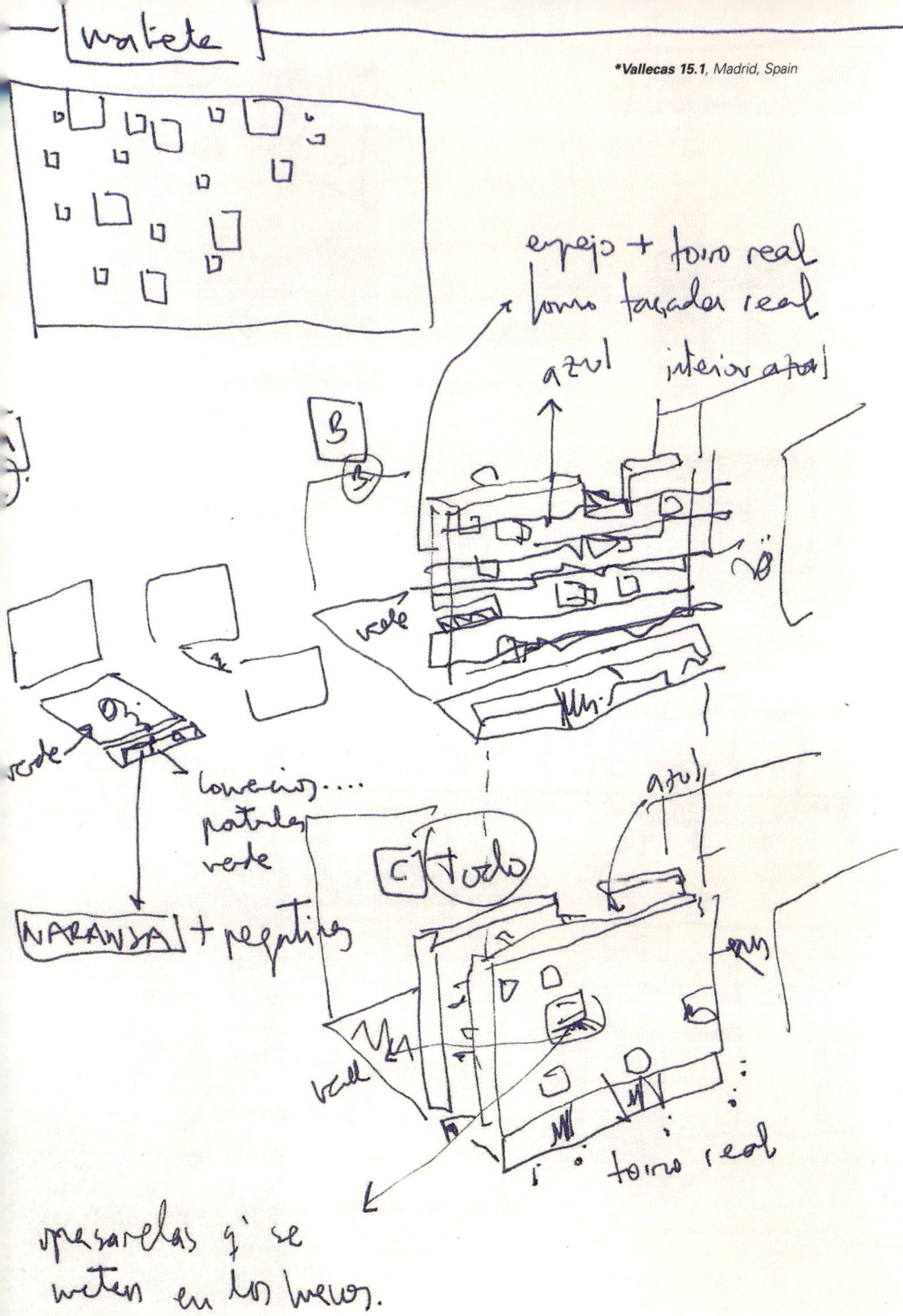

maketa

espejo + toro real
forro fasada real
azul   interior azul

B
vale

comercios....
postales
verde

NARANJA + pegatinas

C todo
azul

vale
toro real

pasarelas q' se
meten en los huecos.

CHA:COL

www.chacol.net

'CHA:COL' (stands for 'Chinmaya Apurva Collaborative'), established in 2006 as a husband-wife multi-disciplinary design collaborative. We have both uploaded sketchbooks. I added text files within each folder to show which notebook belongs to whom.
Our website and contact information is in the email signature at the end of the message.

– Chinmaya's notebook

What is a notebook to you?

To me, a notebook is a natural extension of a mind that **thinks and records visually.** Technology, however advanced, still replicates at some level, the direct connection a sketchbook establishes between the hand and the eye. I think of it as an accurate record of our own growth and change. I often reach for my old sketchbooks to see how I approached a visual situation then and how I think now.

당신에게 수첩이란 무엇인가?

나에게 수첩이란 내 머릿속 생각의 연장선상이다. **시각적으로 생각하고 기록한다.** 기술 또한 어느 정도까지는 손과 눈을 연결 시켜주는 스케치북을 복제할 수 있기는 하다. 나는 스케치북이 우리가 자라나고 변화하는 시간을 정확하게 기록하는 것이라 생각한다. 나는 주로 옛날 스케치북을 보면서 그때는 어떻게 생각했었고 지금은 어떻게 생각했었는지 본다.

Any episodes or memories related to a notebook?

I keep separate sketchbooks, separated by size and paper type for each time I travel. It keeps future organization precise and simple so I can draw in the moment, wherever I feel like on each sketchbook. The separation of sketchbooks helps me connect very directly with memory and association. I often add a small word specific to that place or time, so I can recall why I drew a particular detail or chose an angle.

수첩에 관련된 에피소드가 있다면 들려달라.

나는 여행을 갈 때마다 크기와 종이 종류에 따라 스케치북을 따로 보관한다. 나중에 정확하고 간단하게 정리할 수 있도록 도와주고, 어디서나 그리고 싶은 순간에 각 스케치북에 그릴 수 있도록 해주기 때문이다. 또한 스케치북을 따로 보관함으로써 추억과 그 연결고리를 정확하게 기억해낼 수 있도록 해준다. 주로 그 장소나 시간을 나타내는 간단한 단어를 적어 왜 이 디테일을 그렸고 왜 이 각도에서 그렸는지 생각 날 수 있도록 한다.

When and where do you use your notebook the most?

For work and travel. At work, I keep a small notebook next to my desktop for quick doodles and sketches. For travel, I have a set of notebooks separated by size and flexibility, depending on the size of my backpack.

수첩을 가장 많이 사용하는 공간과 때는?

일과 여행 때 모두 사용한다. 사무실에서는 컴퓨터 옆에 작은 수첩을 두고 간단하게 끄적이거나 스케치해야 할 때 사용한다. 여행을 하면서는 책가방 크기에 따라 크기와 사용도로 나누어진 스케치북을 가지고 다닌다.

What influence does a notebook have in your projects and life as an architect?

While we both pride ourselves in extensive use of digital tools, **I find it hard to imagine our creative process without a physical sketchbook or even a roll of trace around.** Sketching by hand is as much a mental process as it is a physical process. I believe the process of sketching goes beyond recording. It sharpens your thinking as you draw.

수첩은 당신의 작품과 삶에 어떤 영향을 끼치는가?

디지털 툴을 많이 사용하는 것을 자랑스러워 하지만, 주변에 **수첩이나 트레이싱지가 없이 창의적으로 작업한다는 것은 상상조차 할 수 없다.** 손으로 스케치하는 것은 육체적인 과정인 만큼 정신적인 과정도 포함되어 있다. 스케치는 그저 기록하는 것을 넘어선다. 그리는 매 순간 생각을 더 선명하게 만들어준다.

Are there anything else other than a notebook that you use to keep a record of your thoughts and ideas?

We both use digital tools extensively. Mostly for experimentation, organization and post-processing. We have a pen and tablet for digital drawing and a host of software -- one for recording written notes, one for digital doodles and one for organizing scans and images. I experiment with a combination of 3d modeling, vector drawing and raster work.

수첩 외에 자신의 생각을 기록하는 방법과 도구는 무엇이 있는가?

우리는 디지털 툴을 아주 많이 사용한다. 주로 실험적인 프로젝트나 정리, 그리고 후처리 때 사용한다. 펜과 태블릿을 이용해 디지털 그림을 그리고 다양한 소프트웨어를 가지고 있다. 하나는 노트를 적기 위한 것이고, 하나는 그림을 그리기 위한 것이고, 다른 하나는 스캔한 것들과 이미지들을 정리하기 위한 것이다. 3D 모델링과 벡터 드로잉, 그리고 래스터 이미지 모두 사용하여 실험해본다.

I find it hard to imagine our creative process without a physical sketchbook or even a roll of trace around.

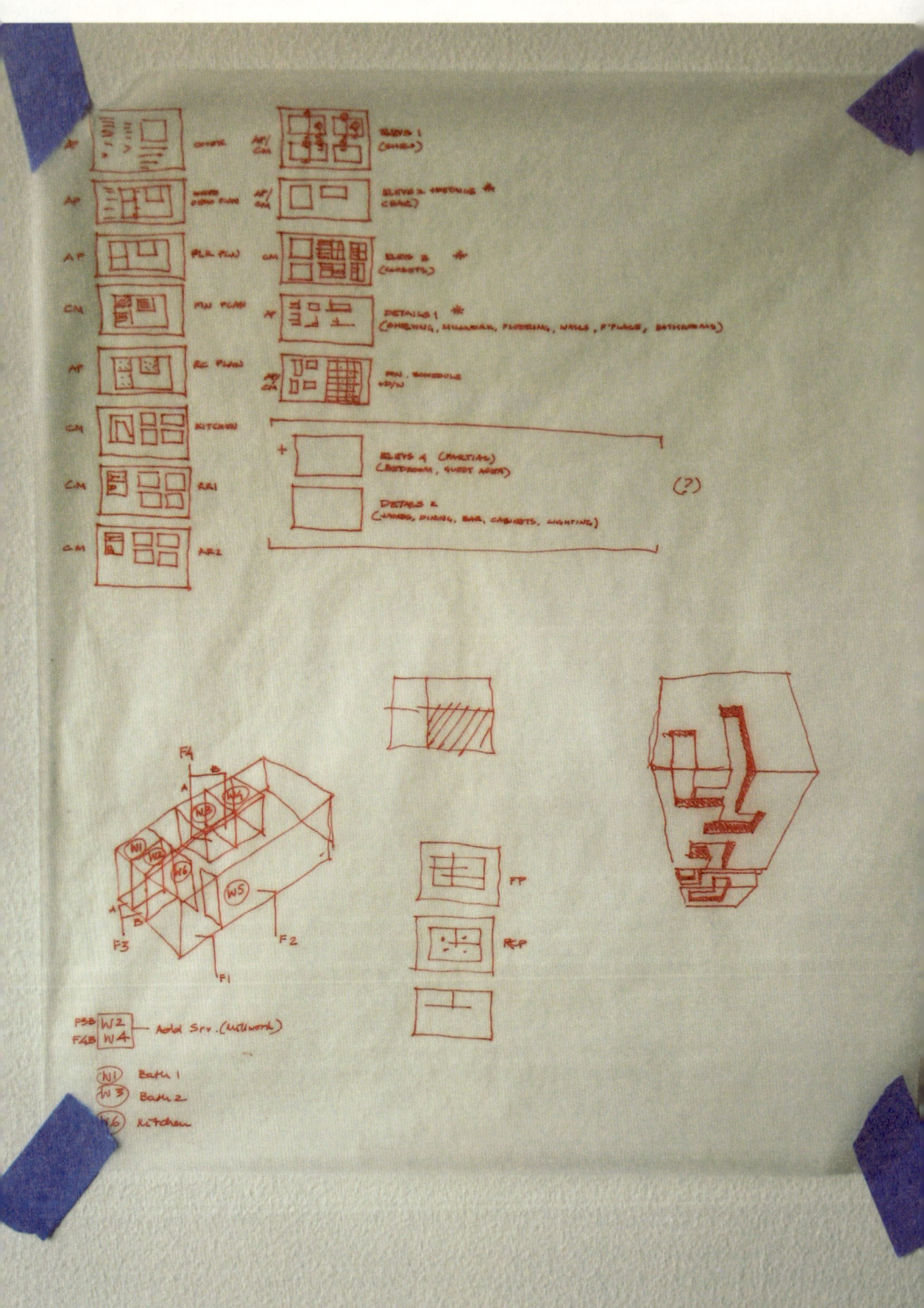

*Bombay

Bombay

*New york

*Paris

© *Santa Ynez*, these notebooks are from Apurva

*Orchha

YOSHIHARA McKEE ARCHITECTS

www.yoshiharamckee.com

Our partnership was established in 1996 with offices in New York and Tokyo. Our projects range in type and size from small residential works to large scale urban interventions. This wide variety of projects has allowed us to work with many types of programs and clients. We have developed innovative uses of under-utilized materials for private residences as well as large scale institutions that chance the urban character of the city.
Sandra McKee is a graduate of the University of Waterloo. She is an adjunct professor at Fordham and Columbia Universities and a visiting critic at architectural and interior design schools including Pratt Institute and City Tech. She is a registered architect in New York State.
Hiroki Yoshihara is a graduate of Nagoya University. He obtained a Ph. D. in physics prior to his study of architecture. He is studying folded paper as a bridge between art and science and its application in architecture. He is registered as a first class architect in Japan.

Interviewee: Sandra Mckee(SM), Hiroki Yoshihara(HY)

What is a notebook to you?

SM: A notebook is full of memories, a journal, a place to jot down notes, reminiscences, ideas.

HY: I have two types, one is a source book of new ideas, the other one is to study ideas; find and investigate solutions.

당신에게 수첩이란 무엇인가?

SM: 많은 기억들이 담겨있는 수첩이자, 노트나 추억, 그리고 아이디어들을 적을 수 있는 것이 수첩이다.

HY: 나는 두 가지의 수첩을 가지고 있다. 하나는 새로운 아이디어들을 담기 위한 것이고, 또 다른 하나는 아이디어들을 발전시키면서 해결책들을 찾고 연구하기 위한 것이다.

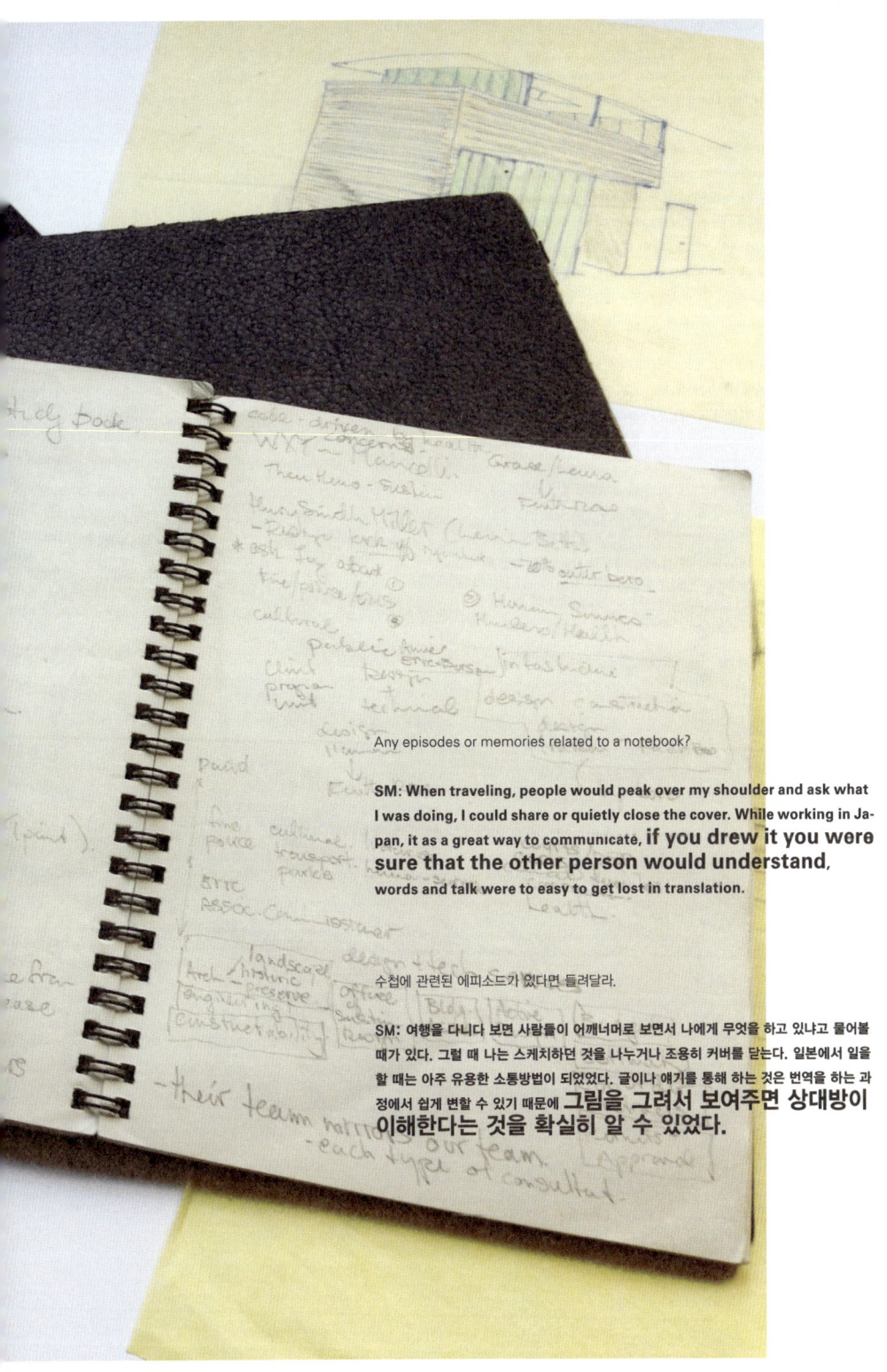

Any episodes or memories related to a notebook?

SM: When traveling, people would peak over my shoulder and ask what I was doing, I could share or quietly close the cover. While working in Japan, it as a great way to communicate, **if you drew it you were sure that the other person would understand,** words and talk were to easy to get lost in translation.

수첩에 관련된 에피소드가 있다면 들려달라.

SM: 여행을 다니다 보면 사람들이 어깨너머로 보면서 나에게 무엇을 하고 있냐고 물어볼 때가 있다. 그럴 때 나는 스케치하던 것을 나누거나 조용히 커버를 닫는다. 일본에서 일을 할 때는 아주 유용한 소통방법이 되었었다. 글이나 얘기를 통해 하는 것은 번역을 하는 과정에서 쉽게 변할 수 있기 때문에 **그림을 그려서 보여주면 상대방이 이해한다는 것을 확실히 알 수 있었다.**

When and where do you use your notebook the most?

SM: In class to "talk visually" with the students, as a travel diary of our trips and at work to develop an idea.

HY: For studying; when I need to find a specific solution, mostly at the office for a project we are working on, but anywhere As a source book for generating new ideas that have not been seen before; when I am very relaxed and inspired by something. Sometimes they are random thoughts. Mostly at home but anywhere, any time.

수첩을 가장 많이 사용하는 공간과 때는?

SM: 학교에서는 학생들과 '시각적으로 소통' 하기 위해 사용하고, 여행을 다니면서 나의 일기장이 되거나, 사무실에서는 아이디어를 발전 시키기 위해 사용한다.

HY: 연구용 – 아무데서나 사용하지만, 주로 사무실에서 프로젝트 작업을 하면서 특정한 해결책이 필요해 연구를 해야할 때 사용한다. 한번도 보지못한 새로운 아이디어의 발상지 – 무언가에 영감을 받고 여유가 있을 때에는 새로운 아이디어들을 담는 것으로 사용한다. 이런 저런 생각이 많을 때도 있다. 주로 집에서 사용하지만, 언제 어디서나 사용 가능하다.

If you drew it you were sure that the other person would understand.

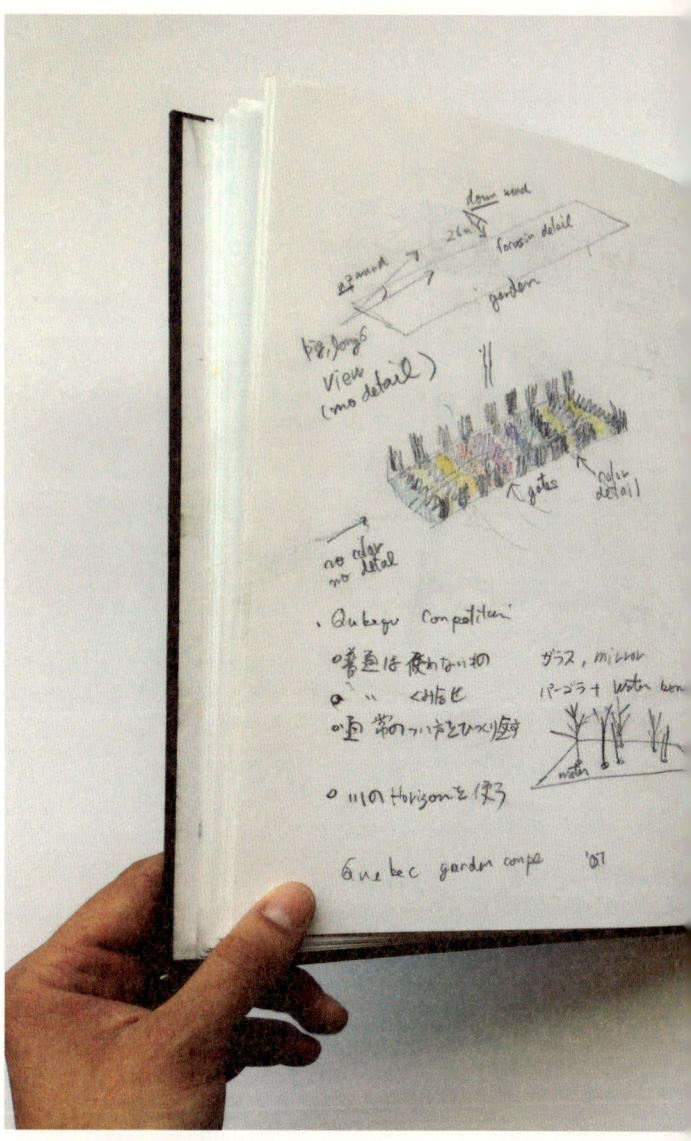

What influence does a notebook have in your projects and life as an architect?

SM: **Provides an overview, a time line**, I still have my sketchbooks form University and travel through them to see how my preoccupations and views have changed.

HY: The problem solving sketchbook is a tool that directly affects our current projects, the source book works future projects and competitions.

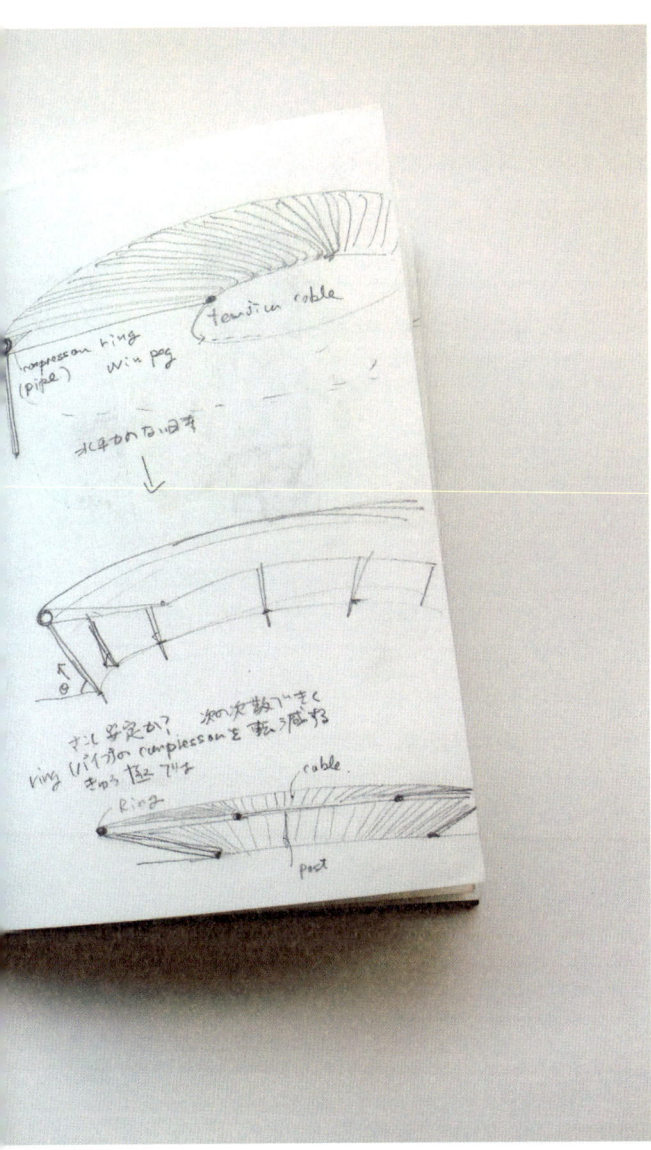

수첩은 당신의 작품과 삶에 어떤 영향을 끼치는가?

SM: 전체적인 개관을 보여주는 하나의 타임라인이다. 대학교 시절과 여행을 다니면서 사용하였던 것들을 아직 가지고 있다. 다시 들여다보면서 나의 선입견과 생각들이 어떻게 바뀌었나 본다.

HY: 문제 해결을 위한 스케치북은 현재 진행 중인 프로젝트들에 직접적인 영향을 끼친다. 새로운 아이디어들을 담는 스케치북은 미래에 있을 프로젝트나 공모전들을 위한 것이다.

I can draw anything sand do it poorly. Actually these sketches almost always turn out to be the best.

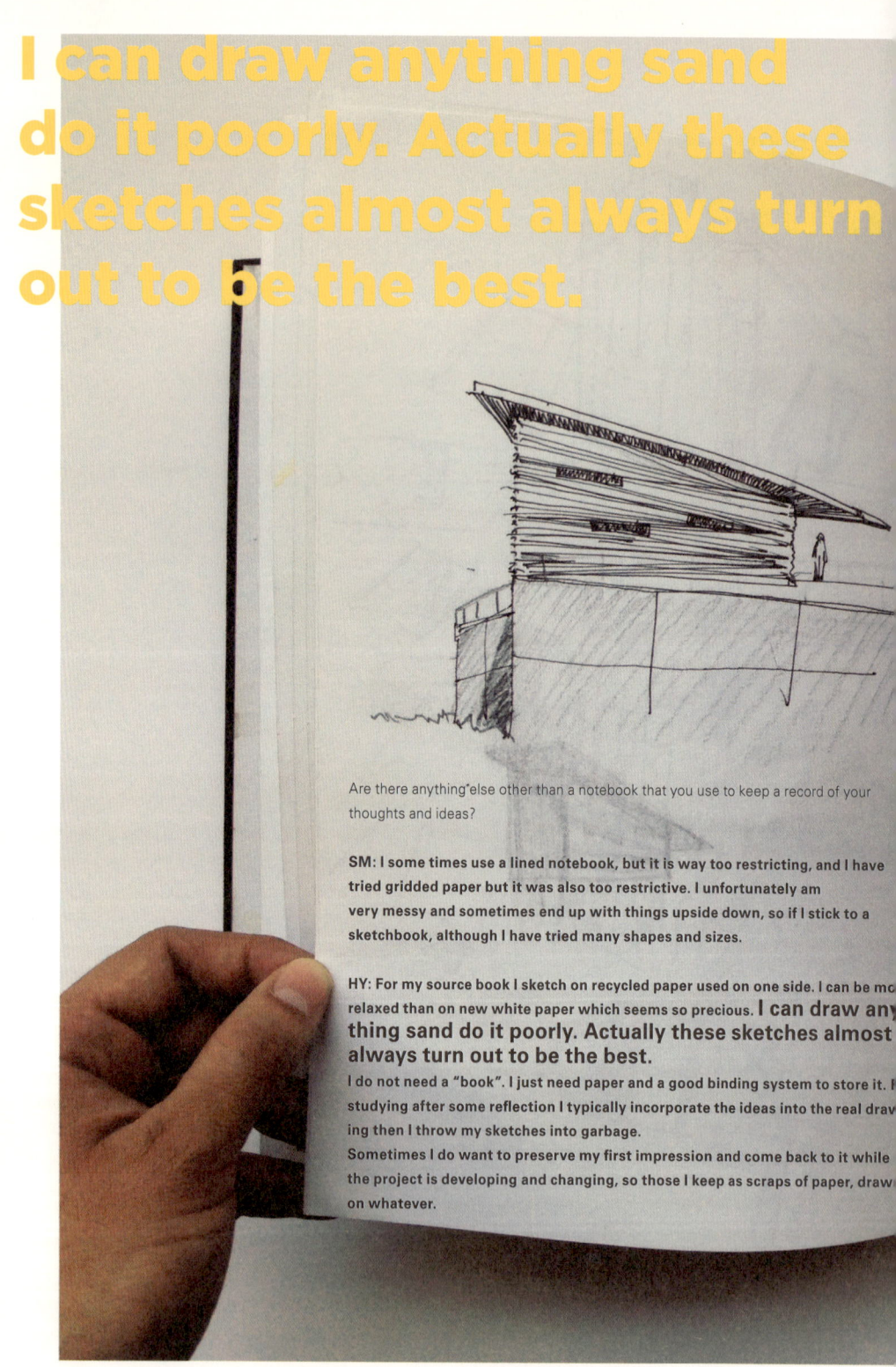

Are there anything else other than a notebook that you use to keep a record of your thoughts and ideas?

SM: I some times use a lined notebook, but it is way too restricting, and I have tried gridded paper but it was also too restrictive. I unfortunately am very messy and sometimes end up with things upside down, so if I stick to a sketchbook, although I have tried many shapes and sizes.

HY: For my source book I sketch on recycled paper used on one side. I can be mo relaxed than on new white paper which seems so precious. **I can draw any thing sand do it poorly. Actually these sketches almost always turn out to be the best.**
I do not need a "book". I just need paper and a good binding system to store it. studying after some reflection I typically incorporate the ideas into the real drawing then I throw my sketches into garbage.
Sometimes I do want to preserve my first impression and come back to it while the project is developing and changing, so those I keep as scraps of paper, draw on whatever.

수첩 외에 자신의 생각을 기록하는 방법과 도구는 무엇이 있는가?

CM: 가끔 줄노트나 모눈종이를 사용할 때가 있는데 너무 제한적이라는 것을 느꼈다. 게다가 너저분한 편이어서 가끔은 거꾸로 그리게 될 때가 있다. 다양한 모양과 크기의 수첩들을 사용해봤지만, 스케치북이 가장 편하다.

HY: 나의 아이디어 북으로는 이면지를 사용한다. 소중해 보이는 새 종이보다 마음 편히 그릴 수 있기 때문이다. **아무거나 대충 그릴 때가 있는데, 주로 이런 스케치들이 가장 좋은 스케치들이 된다.** 나는 '책'이 필요없다. 종이와 제본할 수 있는 것만 있으면 된다. 아이디어를 연구한 후에는 도면에 바로 적용 시킨 후 스케치들은 버린다.
영감을 얻어 그린 첫 스케치는 주로 프로젝트가 발전하고 변할 때에 다시 되돌아볼 수 있도록 간직하고 있는 편이다.

Conlon Loft

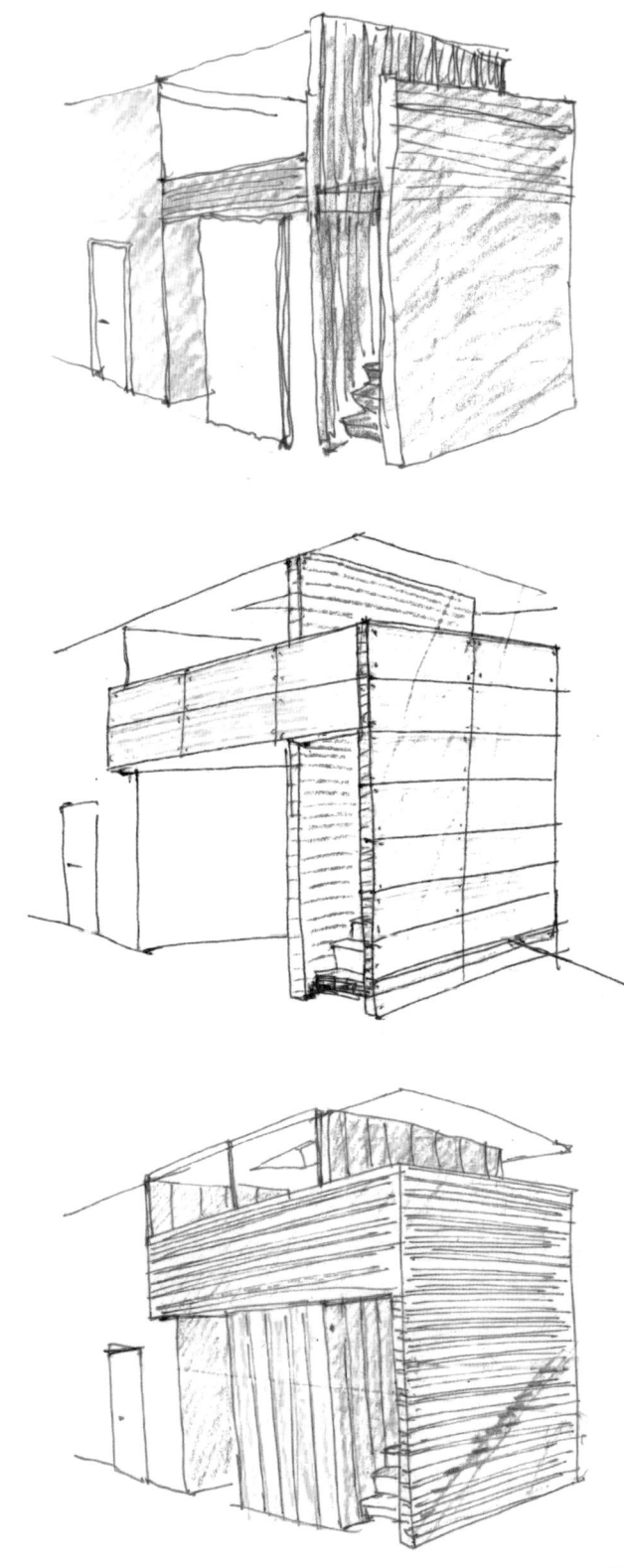

*Posen Loft

*Tea Shop

Gambardellarchitetti

www.gambardellarchitetti.com

Gambardellarchitetti is an handicraft laboratory of architecture design restoration and landscape founded in Naples by Cherubino Gambardella. Chorubino Gambardella, is full professor of architectural design in the Faculty of Architecture Luigi Vanvitelli of the Second University of Naples. Has been involved with theory of the project by writing 10 books and many essays in volumes, catalogs and European magazines. From 1997, Simona Ottieri Gambardella is associated with gambardellarchitetti, and from 2003, she heads the laboratory. They won a series of awards, special mentions, and final selections at the 2003, 2006 and 2009 editions of the Gold Medal for Italian Architecture at the Milan Triennale. These were all for contemporary refurbishment projects of old buildings: the Italian East Africa pavilion at the Overseas Exhibition Site in Naples, a new Kunsthalle for the same site, and the restoration of the Torre dello Ziro on the Amalfi Coast. From 2000 until 2012 they took part in various editions of the Venice Biennale, where their work exhibited the same strongly experimental approach. In 1999 their success in Europan 5 (a Europe-wide international housing design competition for young architects) led them to develop a new system that he used to build an extremely low cost social housing complex at Ancona. In 2002, a hybrid, experimental design proposal saw them successful in another important selection process for approximately 500 social housing units in the northern periphery of Naples, one of the most problematic places anywhere in the Italian urban landscape, between the outlying districts of Chiaiano and Scampia.

It's the place where things are manifested as if it were drawn.

Interviewee: Gambardella

What is a notebook to you? (What does it mean to you?)

A sketchbook is a reserve. Represents a forbidden world where there are no rules. It's an accumulator from which to choose. It's the place where things are manifested as if it were drawn.

당신에게 수첩이란 무엇인가?

스케치북은 특별 구역이다. 어떠한 제약도 없는 금지된 세상을 표현한다. 여러가지 중 고를 수 있는 축적기와도 같고, 그 안에서만큼은 그려지는 대로 이루어지는 공간이다.

Any episodes or memories related to a notebook?

When I began to study architecture gave me fifty notebooks green pocket. I thought that I would have been enough for a long career. Ten years later they were all filled with the things seen, to imagine things, of things which I then built and there was no free space on a white paper.

수첩에 관련된 에피소드가 있다면 들려달라.

건축 공부를 처음 시작하면서 초록색 수첩 50권을 샀었다. 오랜 시간 사용하기 충분하다고 생각했었다. 10년 뒤, 모든 스케치북이 내가 본 것들, 상상했던 것들, 그리고 완공 된 것들로 가득 차서 종이 위에 남는 공간이 없었다.

When and where do you use your notebook the most?

Now I use a lot of my notebooks during traveling or when I do boring meetings at the university.

수첩을 가장 많이 사용하는 공간과 때는?

여행할 때나 대학교에서 재미없는 미팅을 할 때 많이 사용한다.

What influence does a notebook have in your projects and life as an architect?

Drawing is important because it is a storehouse of things that the human brain can not remember. From these things, the project is purified. In a project I do there are many tracks in less than those that design.

수첩은 당신의 작품과 삶에 어떤 영향을 끼치는가?

사람의 머리가 기억할 수 없는 것들을 지장해주기 때문에 스케치는 매우 중요한 것이다. 이 스케치가 프로젝트를 정화시킨다. 프로젝트 디자인을 하는 것보다 옆에서 보조하는 것들이 더 많다.

Are there anything else other than a notebook that you use to keep a record of your thoughts and ideas?

When I'm in the studio I love to draw and compose on large sheets of drawings invention completing or reassembling images taken from the internet. **Just because the design is related to the things I've seen and I like to see right through him.**

수첩 외에 자신의 생각을 기록하는 방법과 도구는 무엇이 있는가?

스튜디오에 있을 때, 인터넷에서 가지고 온 이미지들을 커다란 종이에 스케치하고 합성시키면서 새롭게 재구성하는 것을 좋아한다. **내가 본 것들을 통해 디자인을 하기 때문이다.**

Donner Sorcinelli Architecture

www.donner-sorcinelli.it

Donner Sorcinelli Architecture is an international architectural design office based in Italy, dedicated to create innovative projects.
Founded by architects Luca Donner and Francesca Sorcinelli, the Studio pays particular attention to the theme of sustainable and affordable architecture in all its variants, based on experimentation and research in various fields like Architecture, Planning, Landscape, Interiors and Design.
The research into new architecture typologies and solutions, applied to this themes is represented through various projects developed by the DoSo in a number of countries such as South Korea, USA, Canada, Italy, Finland, UAE and Saudi Arabia.

What is a notebook to you?

It's a way to focus my ideas, thoughts and visualizing schemes and concepts. More of them are plans, sections, perspectives, all together in a sort of bazaar of the architectural process.

당신에게 수첩이란 무엇인가?

아이디어와 생각에 집중할 수 있도록 해주고 계획과 컨셉을 시각화할 수 있도록 도와주는 것이다. 대부분 평단면과 투시도들이 모여서 건축과정의 모음을 만든다.

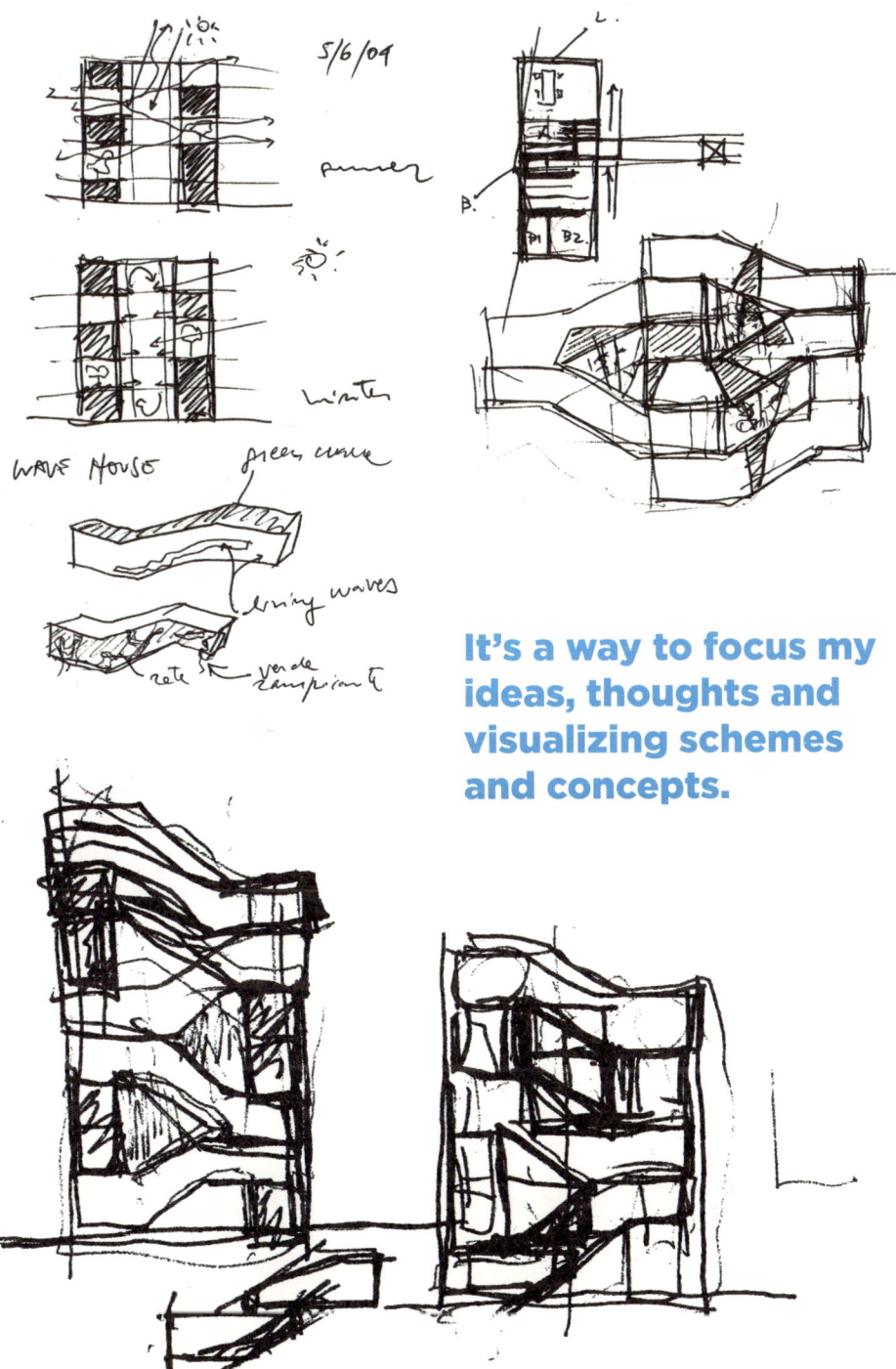

It's a way to focus my ideas, thoughts and visualizing schemes and concepts.

***Wave House**, South Korea*

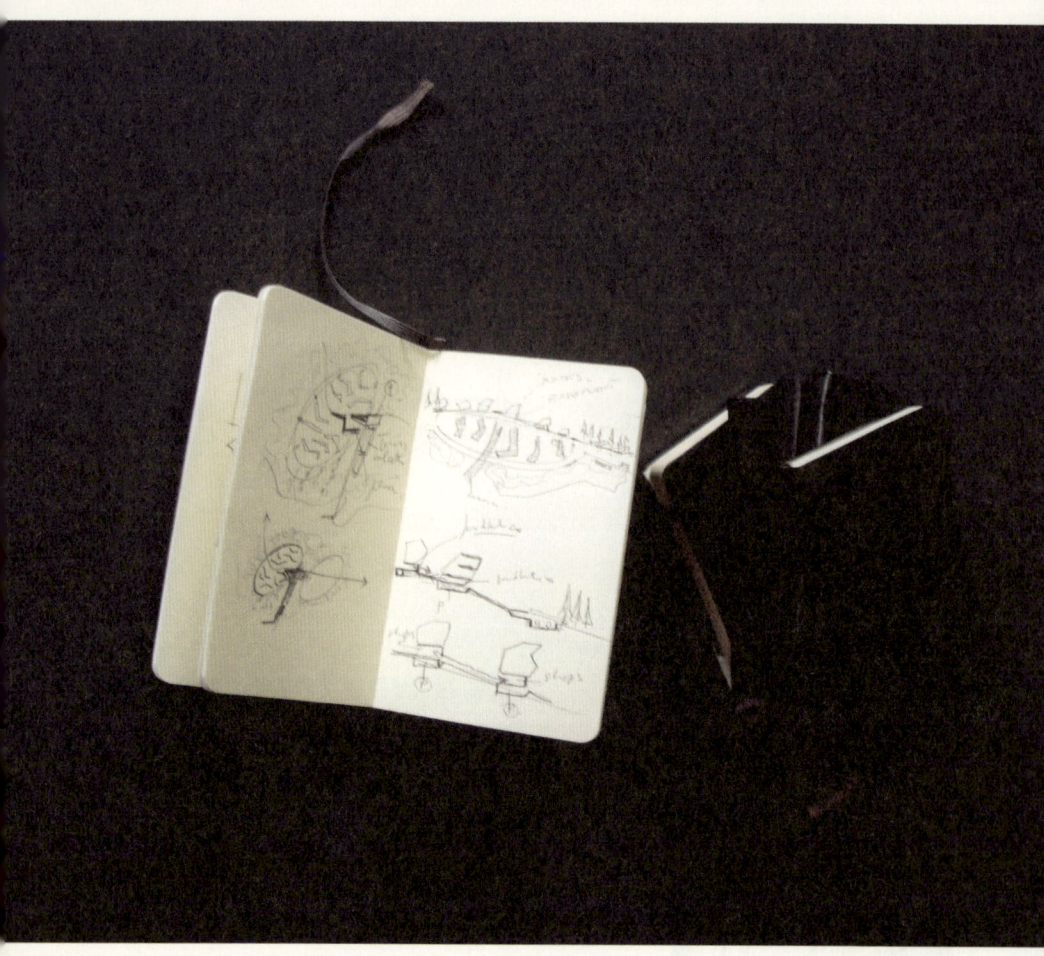

Any episodes or memories related to a notebook?

Episodes are more related when I forget it, so I have to use papers of cafeterias, post it, coasters and everything can be ready to be drawn. These makeshifts becomes attached parts of the notebook, thanks to the pocket at the end of it.

수첩에 관련된 에피소드가 있다면 들려달라.

에피소드들은 대부분 내가 무언가를 잊어버릴 때 생겨난다. 그래서 나는 냅킨이나 메모지, 컵받침과 같이 그릴 수 있는 것은 무엇이든지 사용한다. 이러한 임시방편들은 나중에 수첩에 붙여놓거나 수첩 끝에 있는 포켓 안에 넣어둔다.

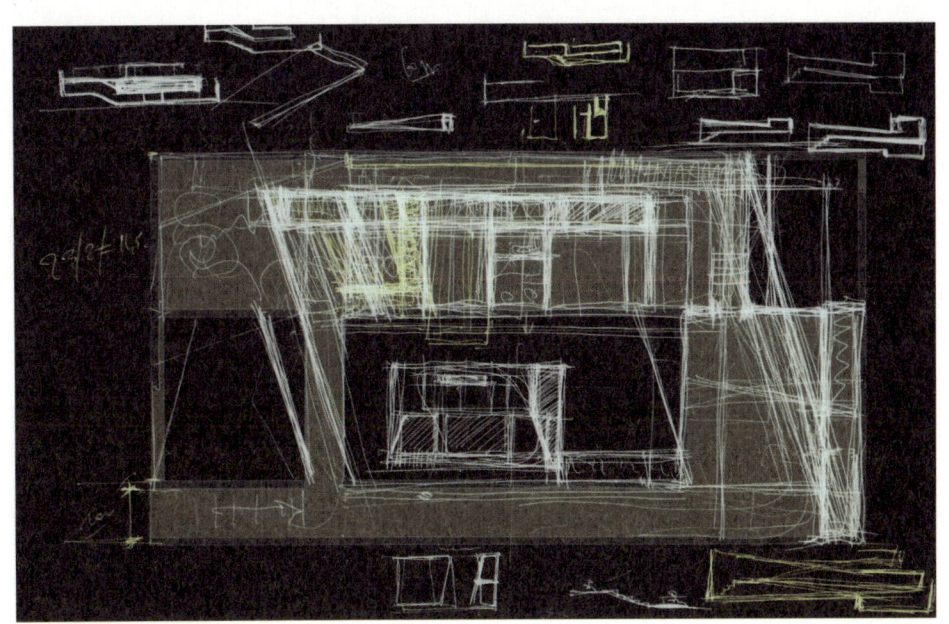

*Breathing House, Saudi Arabia

When and where do you use your notebook the most?

수첩을 가장 많이 사용하는 공간과 때는?

Anytime and anywhere.

언제 어디서나.

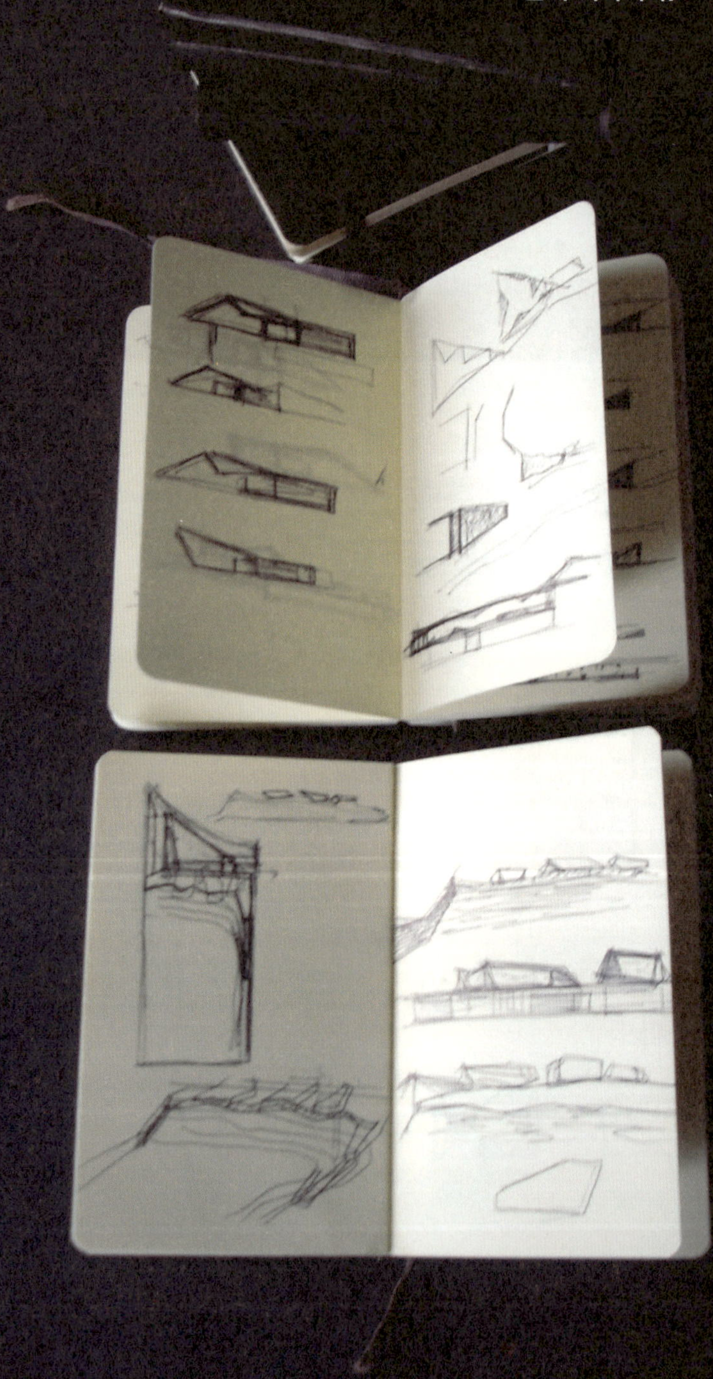

What influence does a notebook have in your projects and life as an architect?

수첩은 당신의 작품과 삶에 어떤 영향을 끼치는가?

It's the way to maintain the route with coherence, driving the process and **sediment ideas for the future.**

일관성 있게 작업할 수 있도록 해주고, 과정을 이끌어나가며, **미래를 위해 아이디어를 축적하는 것이다.**

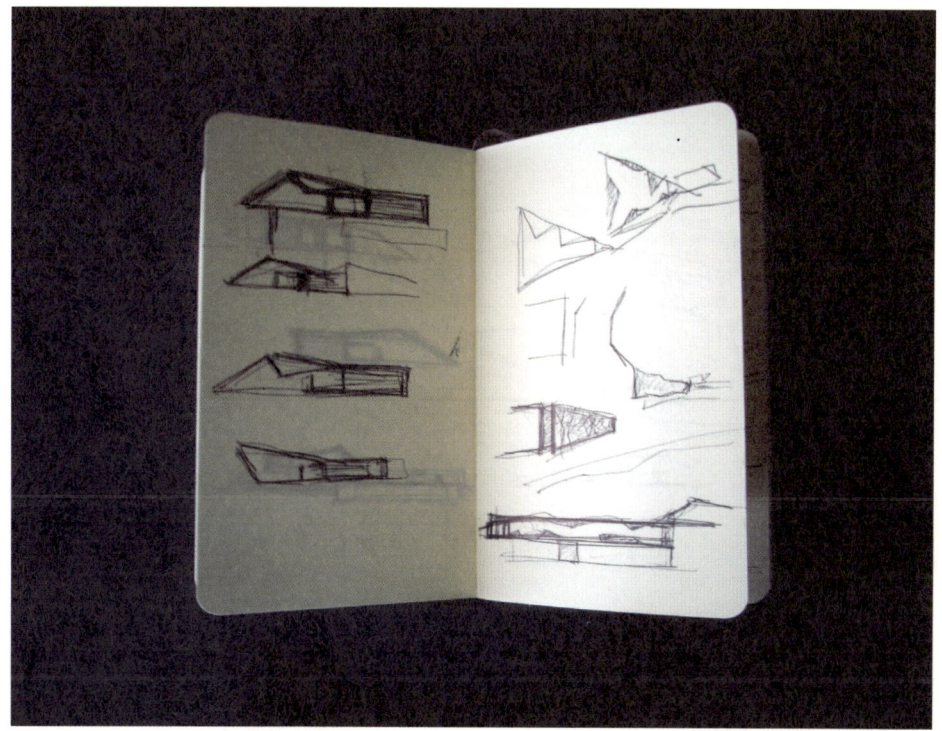

Are there anything else other than a notebook that you use to keep a record of your thoughts and ideas?

수첩 외에 자신의 생각을 기록하는 방법과 도구는 무엇이 있는가?

Everything can be drawn or marked, also the sand on the beach.

바닷가 모래처럼 무엇이든지 그려질 수도 흔적이 남겨질 수도 있다.

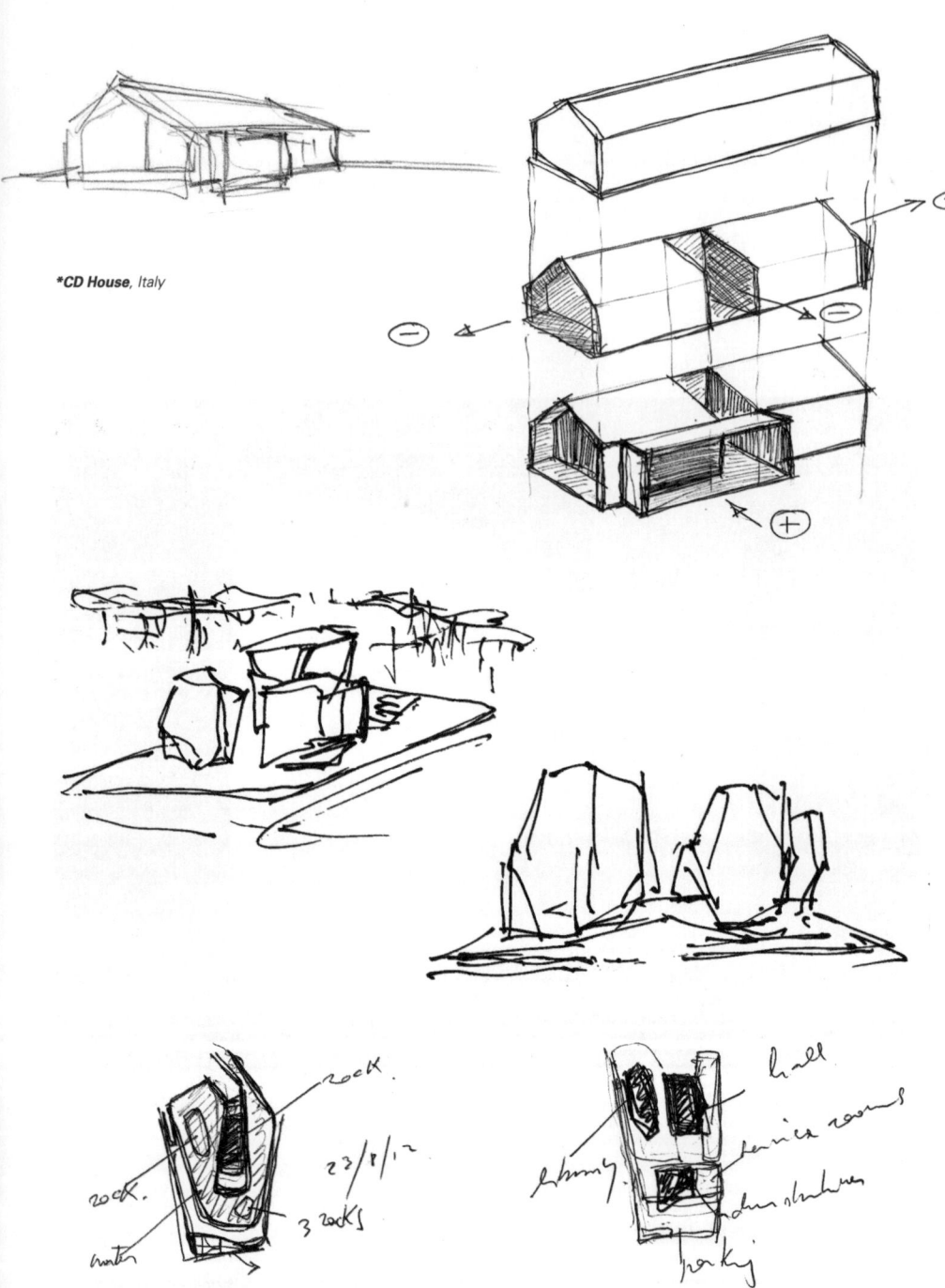

CD House, Italy

Daegu Gosan Public Library, South Korea

New National Museum of Afghanistan

***Butterfly House**, South Korea*

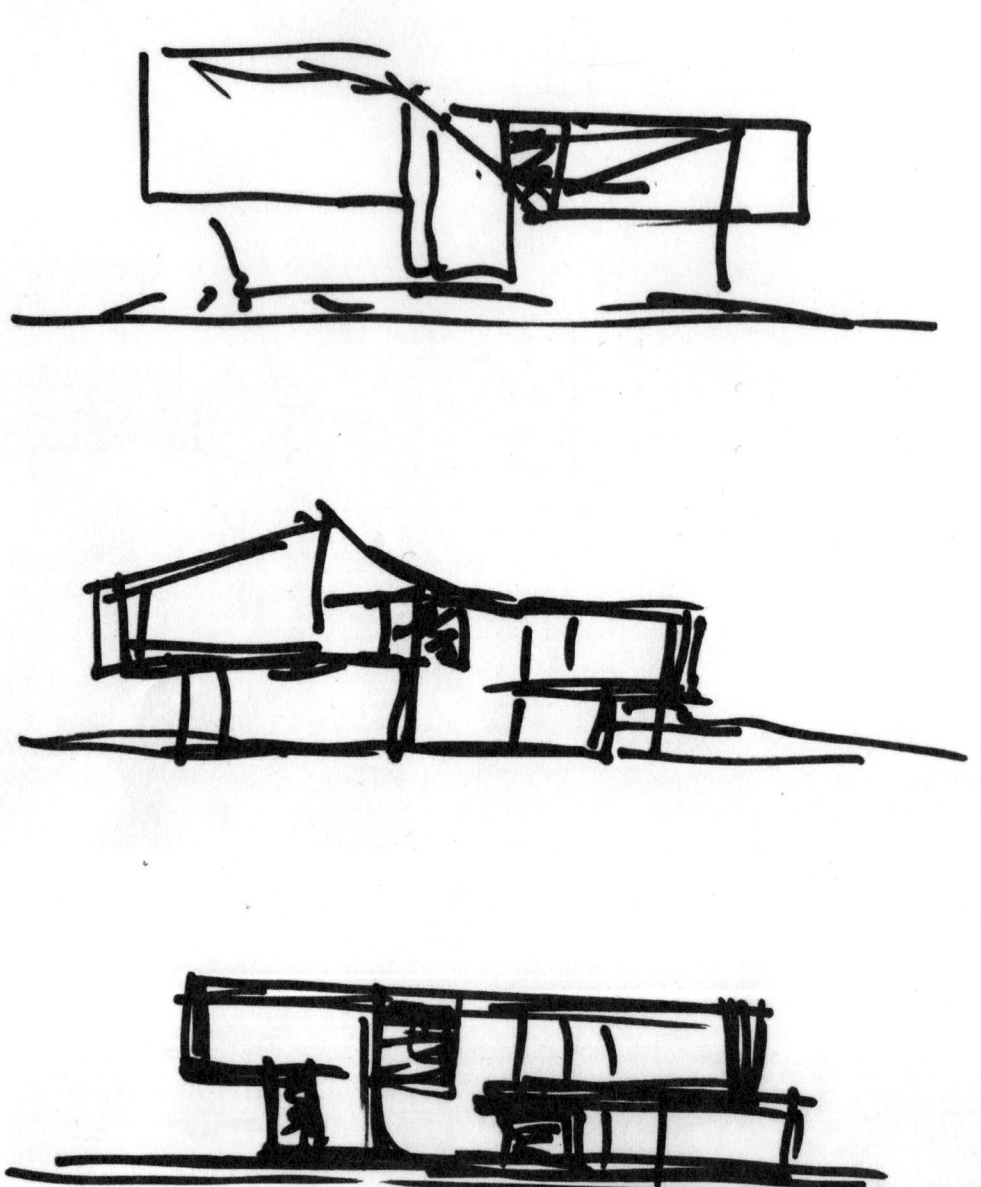

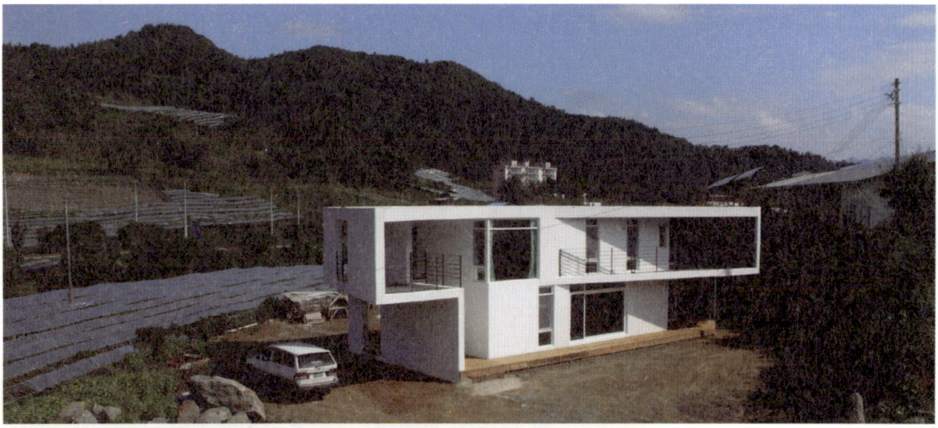

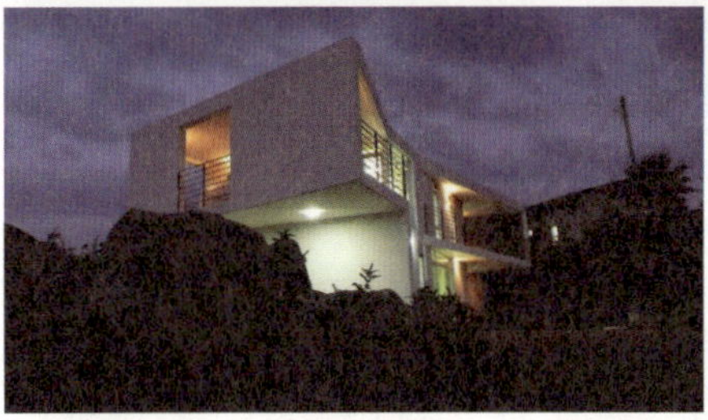

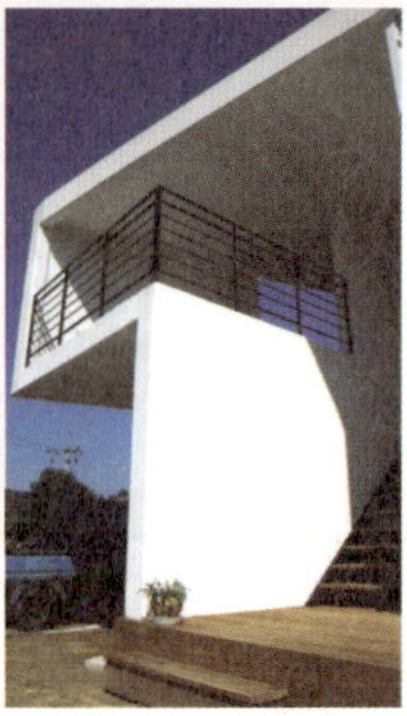

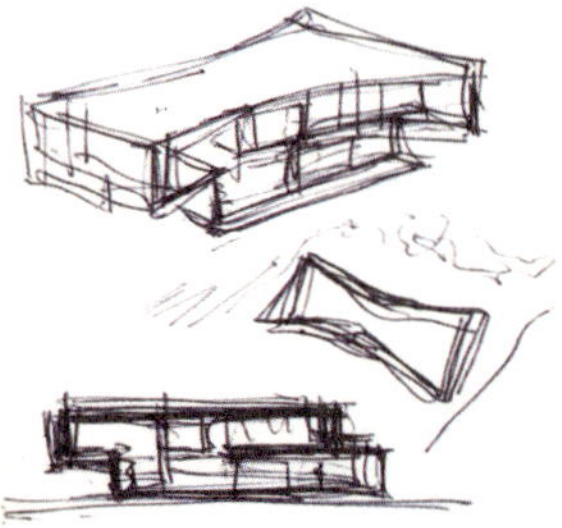

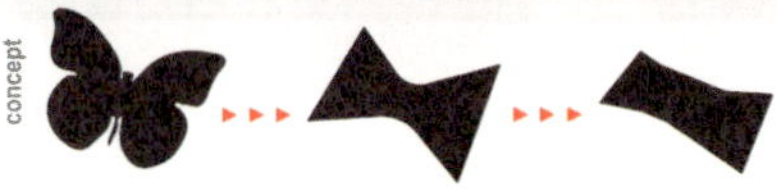
concept

winter prevailing
winds
closed barrier
summer prevailing
breezes
wind collector system
winter
summer
photovoltaic
panels
natural
ventilation
rainwater
collector
sunlight
winter
summer

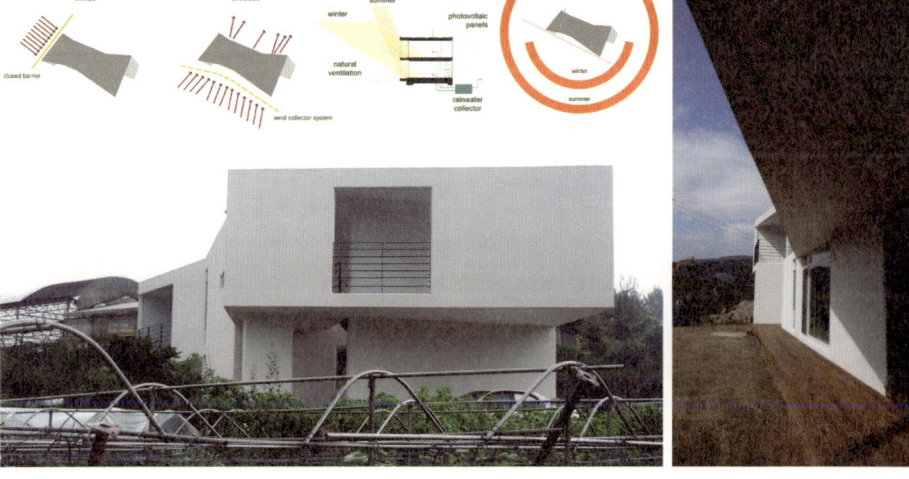

ARHITEKTURA d.o.o.

www.arhitektura-doo.si

My name is Peter Gabrijelčič and I work as a professor at the Faculty of Architecture in Ljubljana, Slovenia, where I have also held the position of the dean for the past 22 years. In addition, I am actively engaged as an architect and together with my two sons, Boštjan and Aleš Gabrijelčič we operate an architecture bureau ARHITEKTURA d.o.o.. I was a guest professor at many universities around the world, I received numerous national and international architecture awards and have, in addition, won many national and international competitions, in recent times also in collaboration with my both sons.

What is a notebook to you?

Free drawing sketch is one of important tools of an architect, with which the architect explores the phenomena of a space and its various atmospheres. **Unlike a painter, where the drawing/picture is the final result of his work, the architect's drawing functions more as his working tool and a method of discovery.** One can only draw what one comprehends. Through observation one grasps the mechanics and psychology of a space and as one sketches his understanding of a space, it becomes a part of one's long-term memory and consciousness. This is the reason why one carries the sketch book at all times, and a mere look at its pages takes one back to the emotional and rational experience of the moment in time and the atmosphere in which the sketches were created.

Unlike a painter, where the drawing & picture is the final result of his work, the architect's drawing functions more as his working tool and a method of discovery.

당신에게 수첩이란 무엇인가?

건축가의 중요한 도구 중 하나가 프리핸드 스케치다. 이를 통해 건축가는 공간과 그 공간이 지니고 있는 다양한 분위기의 현상들을 연구해 볼 수 있다. 그림이 최종 작품이 되는 화가와는 달리, 건축가의 그림은 새로운 것들을 발견하는 과정으로 사용된다. 관찰을 통해 공간의 역학과 심리적인 작용들을 알게 되고, 그 공간을 이해하면서 그릴 때에 그 공간은 이미 자신의 기억과 의식의 일부분이 된다. 그래서 스케치북을 항상 가지고 다니며, 훑어만 봐도 그렸던 순간에 느꼈던 분위기와 감정들을 경험하게 되는 것이다.

Any episodes or memories related to a notebook?

Three years ago we participated in the international competition for four new boulevards in Benghazi, Libya. We only learned of the competition three days prior to its conclusion. As we stopped for dinner on the way back to Slovenia, my colleagues tried persuading me that there is no point in entering the competition due to the lack of time. In addition, we did not have our computers with us. It was then that I picked up a thick marker and drew the concept of a solution on a tablecloth. Three days later we won the competition.

수첩에 관련된 에피소드가 있다면 들려달라.

3년 전에 우리는 리비아 벵가지에 새로운 대로 4곳을 디자인하는 국제 공모전에 참가했었다. 마감 3일 전 공모전에 대해 알게 되었다. 슬로베니아에 돌아가는 길에 저녁을 먹기 위해 식당에 앉아있을 때, 나의 동료들은 시간이 촉박하다며 공모전 참가에 의미가 없다고 했었다. 더군다나 우리는 컴퓨터를 가지고 있지도 않았다. 그 때 나는 두꺼운 마커를 사용해 식탁보에 컨셉트를 그리기 시작했다. 3일 뒤, 우리는 공모전에서 수상했다.

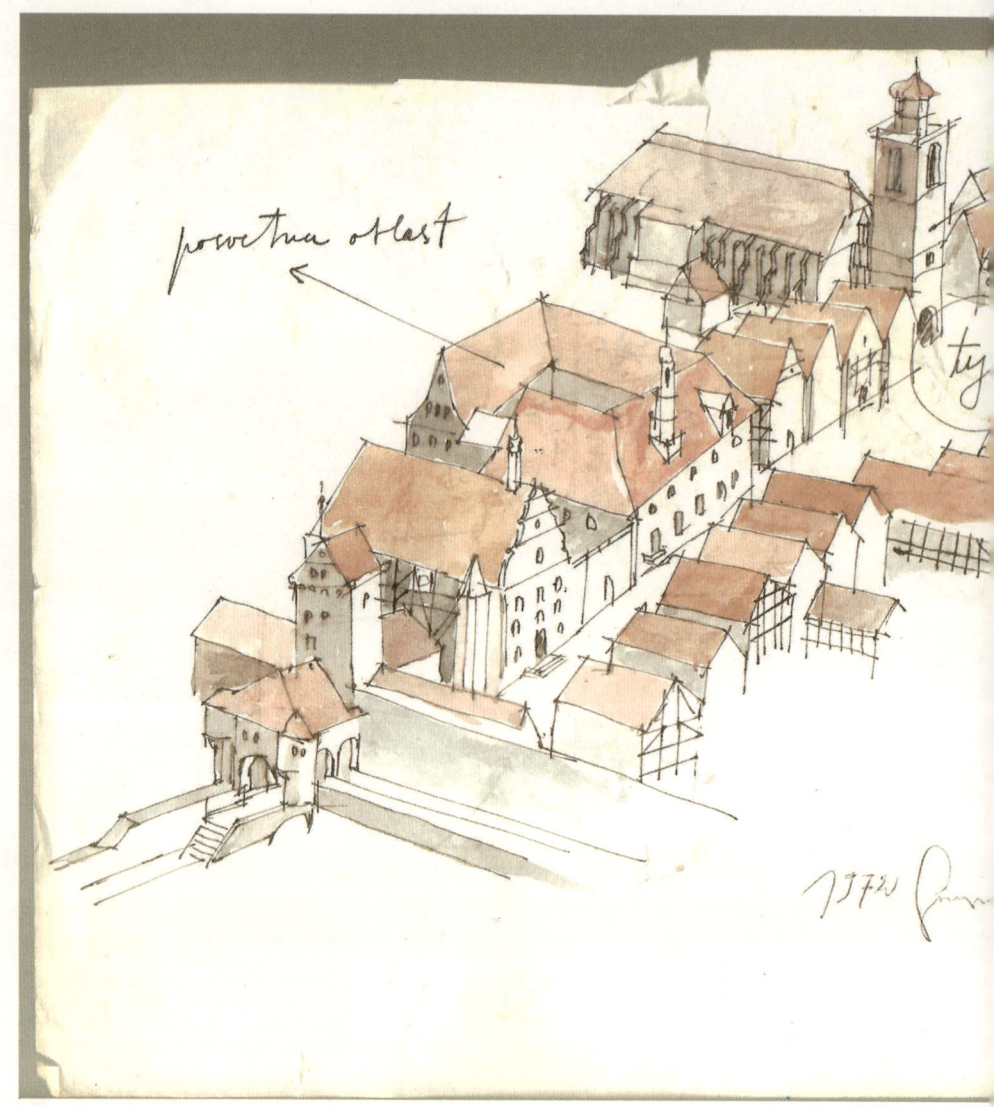

I perceive my hand as a tool though which the subconscious translates into conscious, the intangible into tangible.

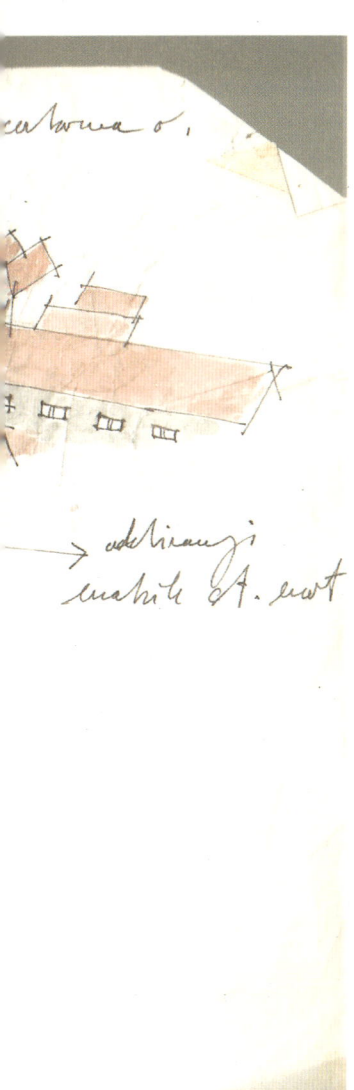

When and where do you use your notebook the most?

I use the notebook mainly as a means of communication with clients. With the help of a drawing I can immediately make every thought tangible, one that has previously only been a part of my or client's perceptions, wishes, intentions. **I perceive my hand as a tool though which the subconscious translates into conscious, the intangible into tangible.** Such a sketch is also easily understandable to a client. This is why, at the end of our session, we often reach a mutual agreement, a visible result, to which the client agrees according to the principle: what you see is what you buy.

수첩을 가장 많이 사용하는 공간과 때는?

나는 수첩을 주로 클라이언트와 소통하는 용도로 사용한다. 그저 클라이언트나 나의 머릿속에 자리 잡고 있던 표상이나 바램, 그리고 의도들을 그림을 통해 표현할 수 있기 때문이다. **나의 손은 무의식을 의식으로, 무형을 유형으로 바꿔주는 도구다.** 또한, 클라이언트가 이해하기 쉽도록 스케치를 한다. 그래서 미팅이 끝날 때쯤이면 우리는 보여지는 결과물에 대해 합의를 한 상태가 된다. 클라이언트는 자신이 본 것을 사는 것이다.

What influence does a notebook have in your projects and life as an architect?

I was once considered to be good at drawing. I was accomplished in the renaissance style of pencil drawing. I used to love travelling in Italy and sketching old towns and palaces. However, this was mannerist drawing, an end in itself. With it one can only draw the architecture of a bygone era. An architecture replete with plastic ornament, reliefs, the play of light and shadow. Later on I noticed that this kind of drawing or sketching only removes me from the creation of contemporary architecture so I rejected it. Today I only use sketching as a working tool (and also to search for an atmosphere).

Are there anything else other than a notebook that you use to keep a record of your thoughts and ideas?

It is my opinion that, in the course of his work, **an architect should use all the possibilities offered by various tools.** I combine sketching with computer drawing, architectural models, photographs, video and sounds. All of the above tools are continuously present on my desk and in my work flow I leap from one form of expression to the next.

수첩은 당신의 작품과 삶에 어떤 영향을 끼치는가?

나는 한때 그림을 잘 그리는 사람으로 알려져 있었다. 특히 르네상스 스타일의 연필화를 잘 그렸다. 이탈리아 여행을 다니며 오래된 마을과 성들을 스케치 하는 것을 좋아했다. 하지만 이것은 그 자체로써 끝이 되는 그림일 뿐이었다. 흘러간 시간속에 남겨진, 플라스틱 장신구나 릴리프, 또는 빛과 그림자만 가득한 건축만 그릴 수 있었던 것이다. 이러한 그림이나 스케치들이 현대적인 건축을 만들 수 없게끔 한다는 것을 깨닫고 다른 방법을 선택하였다. 이제는 디자인 작업을 하거나 분위기를 찾을 때만 스케치를 한다.

수첩 외에 자신의 생각을 기록하는 방법과 도구는 무엇이 있는가?

건축가는 주어진 모든 도구를 사용해 작업을 해야 한다고 생각한다. 나는 스케치뿐만 아니라 컴퓨터, 모형, 사진, 비디오, 그리고 소리도 사용해 작업한다. 이 모든 도구들은 나의 책상 위에 항상 있으며, 다양하게 바꿔가며 작업을 한다.

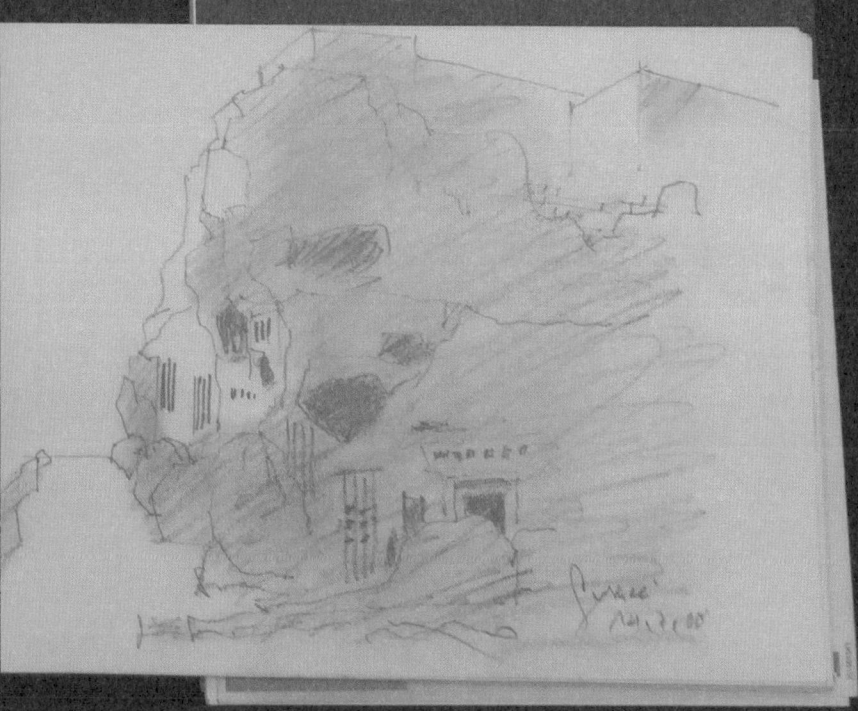

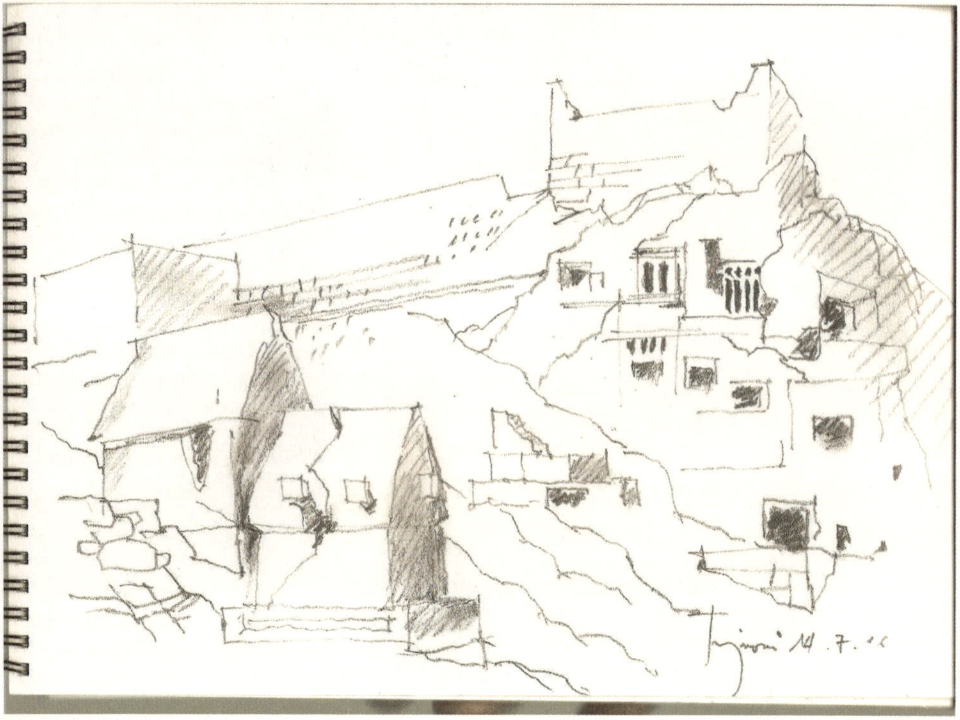

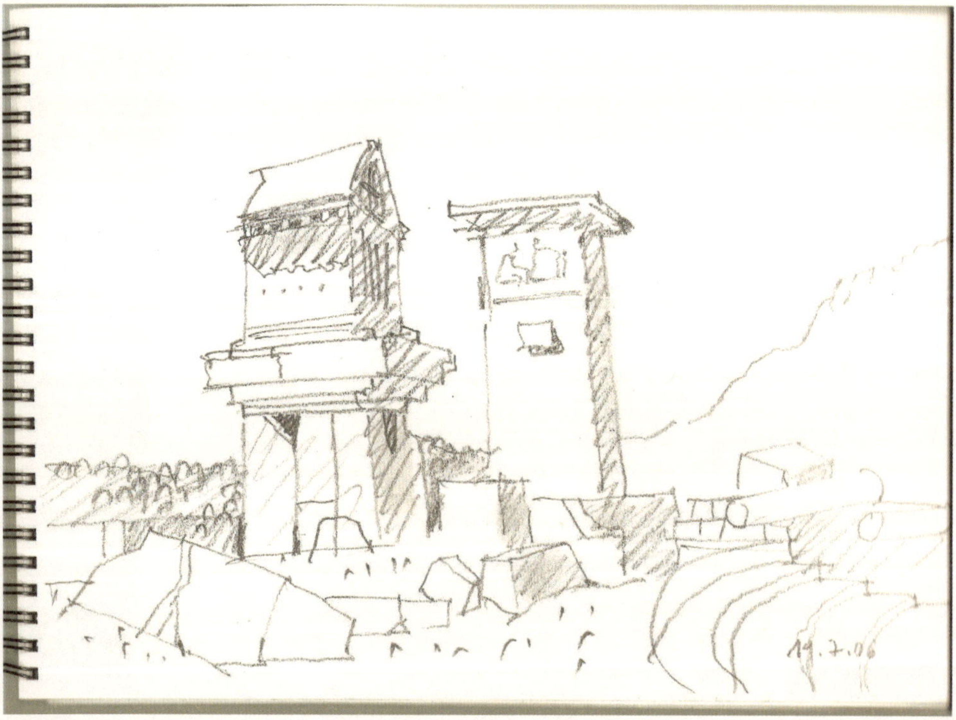
14.7.06

An architect should use all the possibilities offered by various tools.

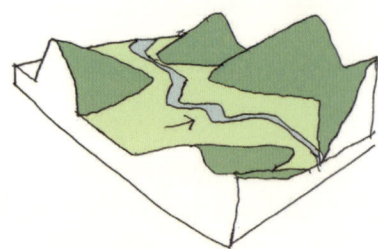

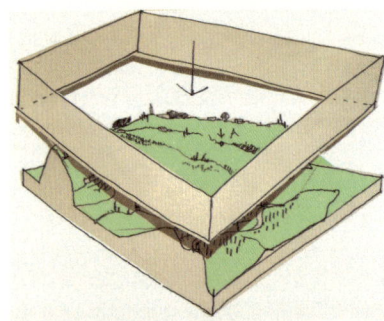
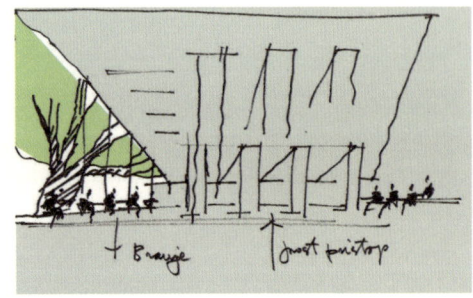

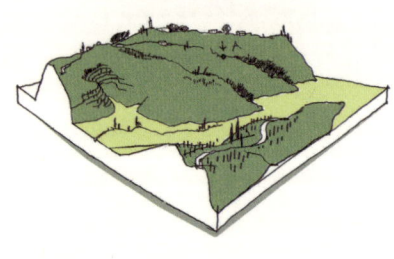

***Studio View**
View from Hybrid Space Lab overlooking Berlin and the river Spree. © Hybrid Space Lab.*

Hybrid Space Lab

www.hybridspacelab.net

Hybrid Space Lab is a think tank and a high-end architecture/design studio.

Hybrid: we combine design fields
Space: our expertise is space.
Lab: we have a long record of highly innovative R&D and design work.

Hybrid Space Lab is an interdisciplinary environment where architects, urbanists, landscape architects, designers and media artists collaborate with soft- and hardware engineers in the development of projects for combined analog and digital, urban, architectural, design and media spaces. The scope of the research and development projects ranges from those on urban games to buildings and architectural interiors, 1:1 industrial design applications and wearables.

Hybrid Space Lab is a R&D and design practice focusing on the hybrid fields that are emerging through the combination and fusion of environments, objects and services today. With an integrated approach to spatial issues Hybrid Space Lab considers objects, services and environments within their networked systems of production, distribution, use and recycling.

Drawings and analog notations are essential within the process of developing ideas, concepts and designs. As our projects are strongly process-oriented we often make notations of processes.

아이디어, 컨셉, 그리고 디자인을 발전시키는 과정에서 스케치와 아날로그식의 표기법은 매우 중요한 부분을 차지한다. 우리 프로젝트는 과정을 중요시 하기 때문에 모든 과정을 표기법으로 나타낸다.

We use communal notation sessions - within the office but also in workshops with external participants. These communal notations are tools to merge analytical investigations with synthetic design processes and to create a networked mental space of interwoven concepts and ideas.

사무실 내에서 사용하는 공용 표기법이 있지만, 외부 사용자들과도 함께 사용한다. 이런 공용 표기법은 분석적인 연구와 통합적인 디자인 과정을 합치는데 도움을 주고 컨셉트와 아이디어가 뒤섞여 있는 네트워크 공간을 만들어준다.

Special for our office are the communal creative tools of **blackboards**. In our studio we use large-scale monumental blackboards to develop ideas together. These are communal creative - sketching, noting and mapping - tools. As we draw and wipe and redraw and reinterpret ... 'new readings' and misinterpretations open up a creative space of unforeseen synapses, nurturing new ideas and innovative understandings. While working together on our blackboards, the most creative moments are the moments of accidental overlaying where new ideas pop-up and evolve.

우리 사무실만의 특이한 점은 바로 다 함께 창의적인 작업을 할 수 있도록 해주는 **칠판**이다. 우리는 함께 아이디어를 발전시킬 때 커다란 칠판을 사용한다. 스케치하고, 표기하고, 매핑할 때 사용하는 창의성 도구다. 그리고 지우고 다시 그리고 재해석할 때에 생겨나는 '새로운 이해'와 오해들은 예측하지 못했던 시냅스를 만들어내어 새로운 아이디어들과 획기적인 생각들을 이끌어낸다. 칠판에서 함께 작업할 때 가장 창의적인 순간은 바로 아이디어들이 실수로 겹쳐졌을 때 새로운 아이디어들이 튀어나와 발전할 때다.

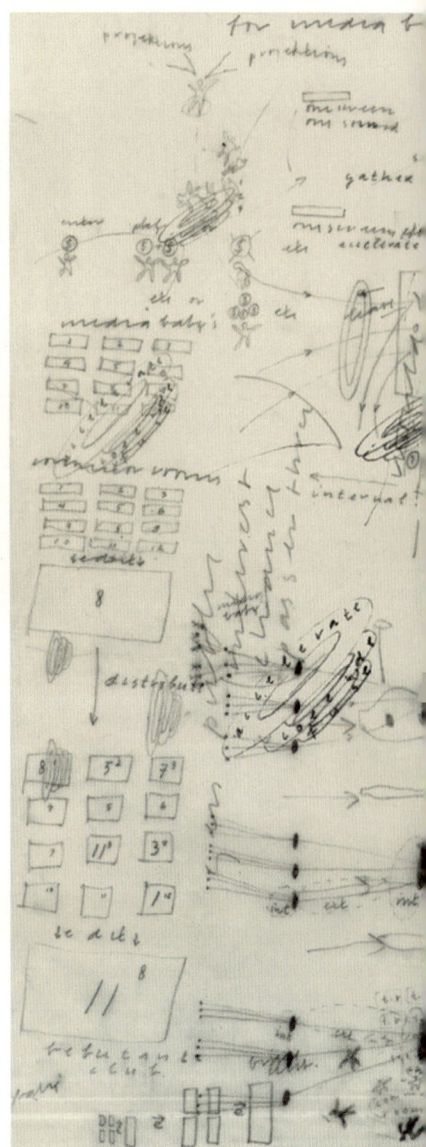

***Blackboard**
In our office we use large-scale blackboards to together develop ideas, concepts and projects. We draw and wipe and re-draw and re-interpret…
© *Hybrid Space Lab.*

***Soft Urbanism**
Notations of dynamic processes in the city. © Hybrid Space Lab.

What influence does a sketchbook have in your projects and life as an architect?

We sketch and make notations within the whole process of project-development. The blackboards are especially helpful as they help develop unexpected synapses, open up minds, unlock ideas, foster creativity.

수첩은 당신의 작품과 삶에 어떤 영향을 끼치는가?

우리는 프로젝트-발전 과정 내내 스케치하고 표기법을 만든다. 칠판은 가장 도움이 되는 도구 중 하나다. 생각지 못했던 시냅스를 발전시키고, 생각을 열어주고, 아이디어들이 떠오르게 하면서 창의력을 키워주기 때문이다.

***Workshop Mappings**
Notations from the workshop conducted by Hybrid Space Lab within the framework of "Torino 2008 World Design Capital" on the theme "Open and Safe Places" attended by 40 international young professionals. © Hybrid Space Lab.

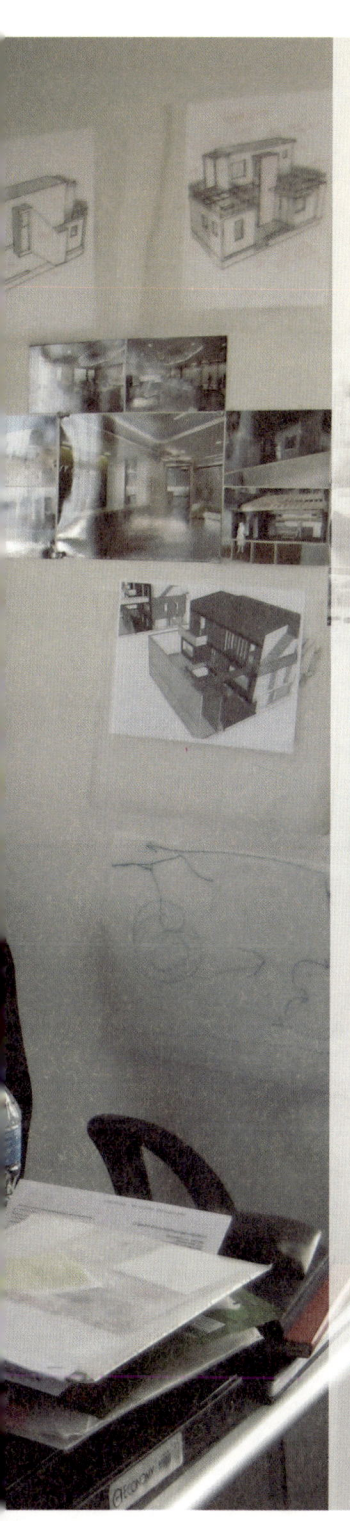

eu.k Architects

www.eukarchitects.com

eu.k architects was founded by architects Jungwoo Ji who has been teaching at Iowa State University and Kyoung Eun Kwon who is an adjunct professor at Ewha Womens University and is doing local and international practice in Korea, China, Middle East, and US. eu.k is focusing on the Boundary and Gradation to provide a frame for people to fill their stories, memories, and dream in since the name of eu.k means 'Good Boundary'.

the studio itself becomes a large 3-dimensional notebook.

What is a notebook to you?

Our notebooks represent our office (studio) itself. When a project begins, numerous sketches begin to fill up the wall. There are various kinds of sketches, from conceptual to furniture details, from presentation sequence to the architectural form; in between them are photos and study models. As we develop the project by discussing the ideas while walking through the spaces, **the studio itself becomes a large 3-dimensional notebook.** In other words, the project process itself is the organic notebook. Moreover, when we go out for meetings or lecture at schools, we use 'Satellite notebooks' which are actually regular notebooks. All the sketches and notes made on these notebooks are connected and supplemented to the 'Core notebook' which is our studio.

당신에게 수첩이란 무엇인가?

우리의 수첩은 사무실 (스튜디오) 그 자체이다. 프로젝트가 발생하면 수많은 스케치들이 벽에 붙기 시작한다. 추상적인 개념스케치에서부터 가구 디테일의 스케치까지. 프리젠테이션 시퀀스부터 건축형태까지. 그 사이사이를 사진과 스터디 모델들이 개입한다. 그 공간을 산책하고, 토의하며 프로젝트를 진행해 나가기에 **스튜디오 자체가 큰 입체 수첩이다.** 즉, 프로젝트의 진행 과정이 유기적인 수첩이다. 아울러 사이트에 나가거나 외부 미팅을 나가거나 학교에서 강의를 할때는 '위성 수첩(Satellite notebooks)'을 사용한다. 그 수첩에 기록되는 스케치 혹은 노트들은 '본체 수첩(the Core notebook)'에 접속되고 보완을 이룬다.

Any episodes or memories related to a notebook?

We sketch ideas related to the projects at home as well. Those usually end up on the wall beside my desk as well. So ultimately, the wall at my house becomes a 'satellite notebook' as well. One day, **my seven year old son** was staring at one of the sketches on the wall and said, "I am going to draw that." Then, he grabbed a trace paper and started to trace over the sketch. After a few months, he looked at a model, drew it on a piece of paper, and pasted a tree made out of real wood. He has made a hybrid expression by combining a 2D sketch made from 3D model and a 3D object together in one. **Even he is sketching in organic methods.**

수첩에 관련된 에피소드가 있다면 들려달라.

집에서도 물론 프로젝트와 관련된 아이디어 스케치를 하곤 한다. 그것들중 역시 주로 내 자리 벽에 붙인다. 그러므로 집 벽 또한 '위성 수첩' 의 하나입니다. 어느날은 그중의 한 스케치를 물끄러미 보던 7살 아들 이 '저걸 그리겠어요' 하며 그 위에 트레이싱지를 덮고 스케치를 형태를 따라 그리는 것이다. 그러더니 몇 개월 후에는 건축 모형을 보며, 다시 자기만의 언어로 종이 위에 스케치를 하고, 그 위에 나무를 잘라서 붙입니다. 즉, 3D, 모델을 보고 2D 스케치와 3D 오브젝트를 결합한 하이브리드 표현을 하는 것이다. 심지어 아이도 유기적인 방법으로 스케치를 남긴다.

When and where do you use your notebook the most?

The most times I use a so called notebook is during architecture studio critiques at schools. In order to give feedback based on the students' materials, I need to remember their project process and my comments for every critique I have with them. So, I note everything down inside the notebook.
However, since we always sketch with various design materials at the office, especially together at a large conference table, the large sketchbook is always short on pages.

수첩을 가장 많이 사용하는 공간과 때는?

일반적인 의미의 수첩을 가장 많이 사용하는 때는 학교 건축스튜디오 크리틱을 할 때이다. 여러 학생의 작업 진행을 기억하고 자신의 코멘트도 기록에 남겨야 다음 크리틱 때 더 발전된 안을 바탕으로 이야기 할 수 있으므로 수첩에 꼼꼼히 기록해놓는다.
그러나 사무실에서 늘 스케치를 하고 있고 특히, 큰 회의테이블에서 여러 디자인 머티리얼들과 함께 스케치를 하기때문에, 이 거대한 수첩은 늘 페이지를 넘겨간다.

What influence does a notebook have in your projects and life as an architect?

A sketch is an important factor in communicating ideas and coming up with a solution. A project called 'Gu-san-seo-ga' (library village in Gusan town) was a collaboration project with another office. During the process, many sketches were used to communicate with each other. In some cases, the partners of each office came up with one sketch together. Sometimes a physical notebook can touch or simulate someone. A notebook filled with the process of the students' work leaves a deep impression to the students.

수첩은 당신의 작품과 삶에 어떤 영향을 끼치는가?

다양한 의견을 교환하고 합의하고 전달하는 데 스케치는 중요한 역할을 한다. '구산서가 龜山書街'라고 하는 프로젝트는 다른 사무실과 협력하여 진행하였는데, 그 과정에서 수많은 스케치들이 의사소통을 도와주었다. 심지어는 두 사무실의 소장들이 협력하여 하나의 스케치를 만들기도 했다.
때때로 물리적인 수첩은 타인에게 감동 혹은 자극을 주기도 한다. 학생들의 작업 과정을 꼼꼼하게 기록한 수첩은 학생들에게 깊은 인상으로 각인된다.

Are there anything else other than a notebook that you use to keep a record of your thoughts and ideas?

Other than the traditional notebook, I recorded my thoughts in various mediums, such as books, writings, blog, and webzines. Recently, I began to use twitter, facebook, and other SNS mediums to leave thoughts and communicate with others regarding them. **All these move 'together.'**

수첩 외에 자신의 생각을 기록하는 방법과 도구는 무엇이 있는가?

정통적인 수첩 이외에 책, 원고, 블로그, 웹진 등을 이용해서 생각을 기록해왔고 최근에는 트위터, 페이스북 등 SNS를 통해서 기록을 남기고 관련된 의견들을 소통하기도 한다. **이 모든 것이 '함께' 이용되고 움직인다.**

[AFTER THE PIN-UP 04/12/13]
- Eli "Seattle Urban Bicycles Archive"
  - Should Finalize Floor Plans
  - Figure out the bike ramp angle in section & Model
  - Building Model w/ Structure frames, stairs, ramp, elevators, slabs, and skin
  - Define the relationship between Sound and by bicycles
  - Retail model and drawings for movable parts

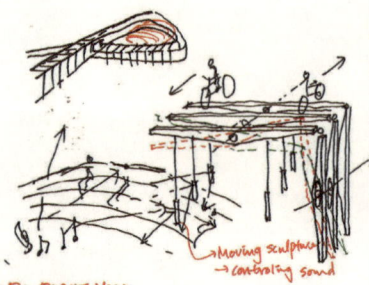

※ DO RIGHT NOW,,
- Plans & 1/8" Model (Basswood)

[AFTER THE PIN-UP 04/12/13]
- Brady "Seattle Urban Mobile Archive"
  - Finalize Site Plan and then Floor Plans
  - Develop Structure for frame
  - Develop Mobile Booths design and movable parts
  - Elevation study from sculpture park.
  - Diagrams for different uses and events

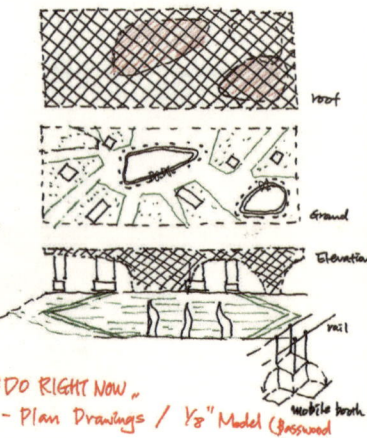

※ DO RIGHT NOW,,
- Plan Drawings / 1/8" Model (Basswood / wire)

[[AFTER THE PIN-UP 04/12/13]
- Rachel "Seattle Touch Archive?"

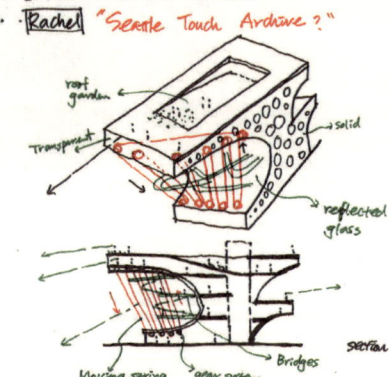

- Develop Section Drawing
- Finalize Floor Plans
- Develop moving system as model & drawing hologram
- Design elevation & material
- Structural solution

※ DO RIGHT NOW,,
- Section and 1/8" Building Model (Basswood w/ wire)

[AFTER THE PIN-UP 04/12/13]
- Stanley "Seattle Urban ▢ Archive"
  - Finalize Floor plans w/ relocating elevator
  - Section Drawing w/ Actual building height
  - Structure model to show space frames
  - 1/8" Building model to show slab level change & building skin
  - Diagrams for circulation, programs, and views
  - Relocate corner program for quiet room

- Detail model of Skylight w/ Movable mirrors

※ DO RIGHT NOW,,
- 1/8" Building Model (Basswood w/ wire)

4/5 OVER the weekend ARCH 202

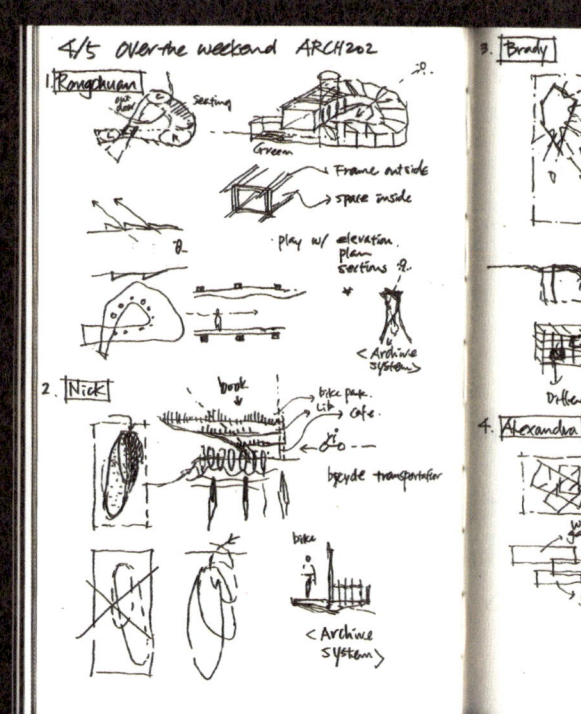

1. Rongchuan
2. Nick
3. Brady
4. Alexandra

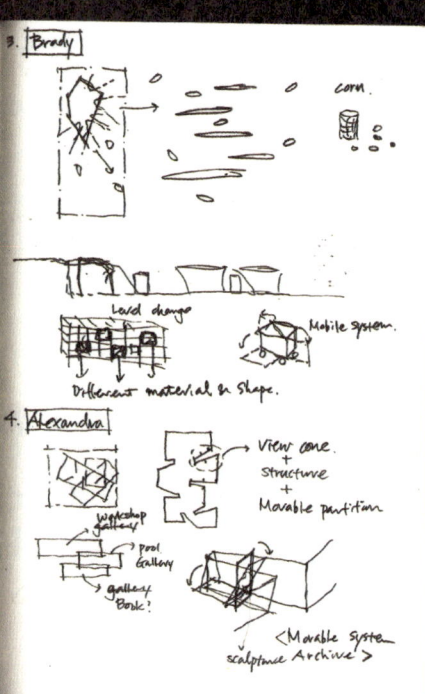

[AFTER THE PIN-UP 4/12/13]

Zachary — "Urban Homeless Archive"
- Section Study
- Secondary Skin shape & area
- Movable furniture detail and connect to secondary skin
- On floor plan: one more stair using outside
- Diagrams for Skin & Furniture
- "DO RIGHT NOW"

1/8" scale physical model (Bass Wood) w/ structure, mullions, secondary skin, movable parts, floor openings, stairs, space frame

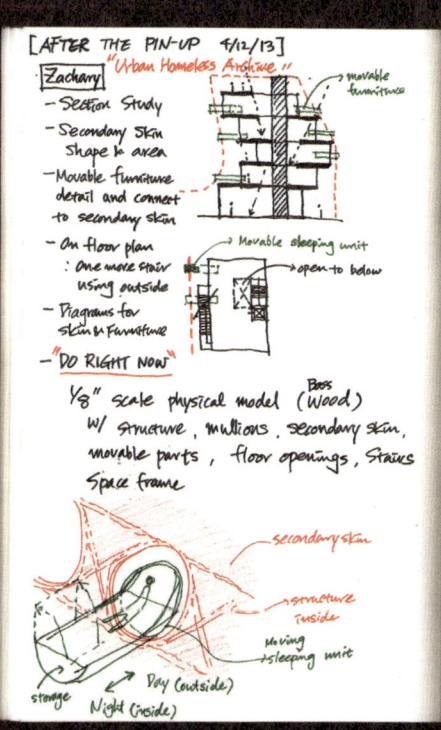

[AFTER THE PIN-UP 4/12/17]

Taylor — "Urban Texture Archive?"
- Plans modification
- Structure Drawing: Axonometric w/ showing movable skin
- Structure Detail: how to move partition wall and curtain
- Diagrams for Space changes w/ Movable parts: Multiple diagrams to show different events or situations
- Building material relation study
- Do Right Now: 1/8" scale physical model (Bass wood) w/ Structure, Elevation, Movable parts, Vertical circulations... Frame...

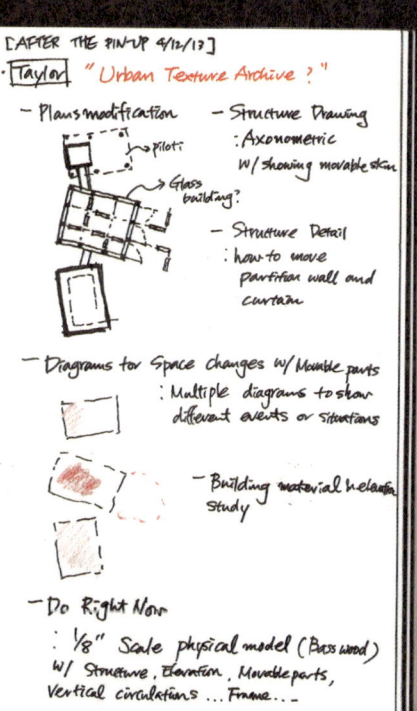

3/29   ARCH 202

Alexandra : - Swimming Pool + Gravity

Rachell:  ≡ sound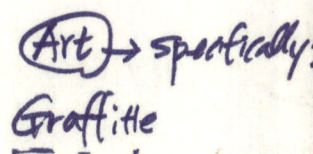
using this

Eli: 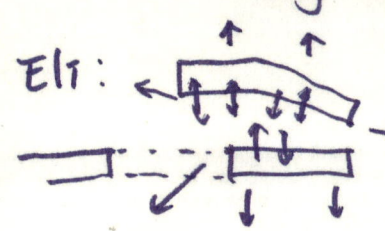 → Bike path
Seattle Music performance
Recreation.

Muhammad

   Sound
Water
Gravity
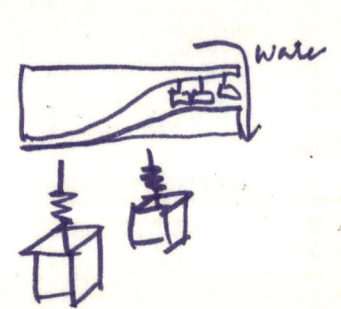

Zach

   Homeless people

- **Brandon**

  too direct.
  photo Gallery.
  Taking Pictures
  ★ Camera obscura research the!

- **Ronchuan**

  Too symetrical
  Too symbol.
  light.   Site analysis first.

- **Brady**

  specify texture
  → using inbetween space
  for reading.

- **Chunhui**

  → 3D → Track
  context

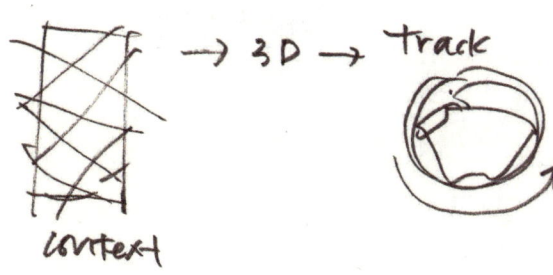

exist parking
Parking
parking at B2
urban fabric
plaza
alley
tube
buffer
main
reno + ent.
NEW urban fabric
B1 sunken & performance
various open space
forest
kidz + eco
community + event
reading + cafe
Main ent.
Sub ent.
ACCESS to neighbors
master plan
solid & void

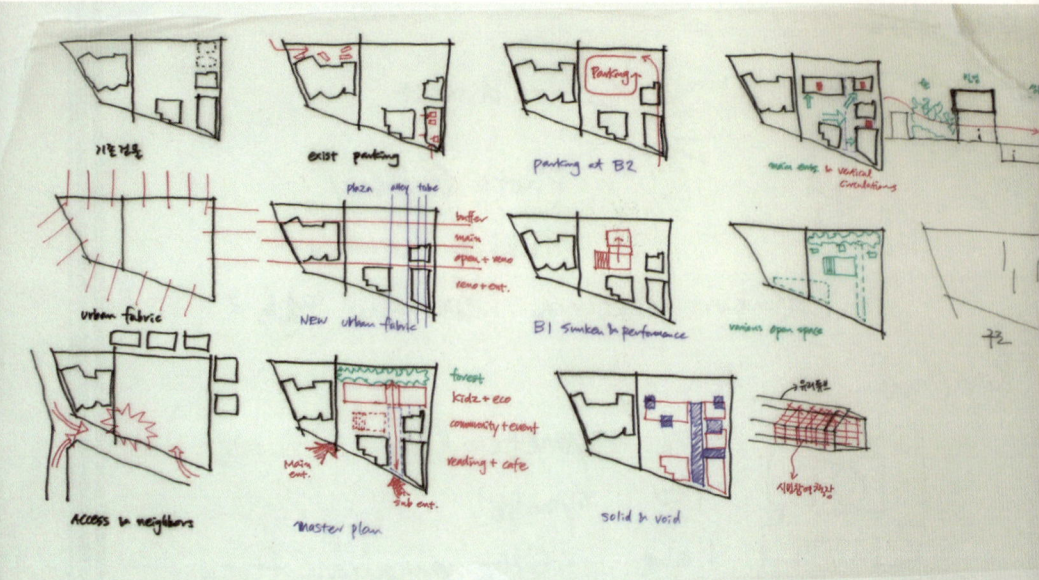

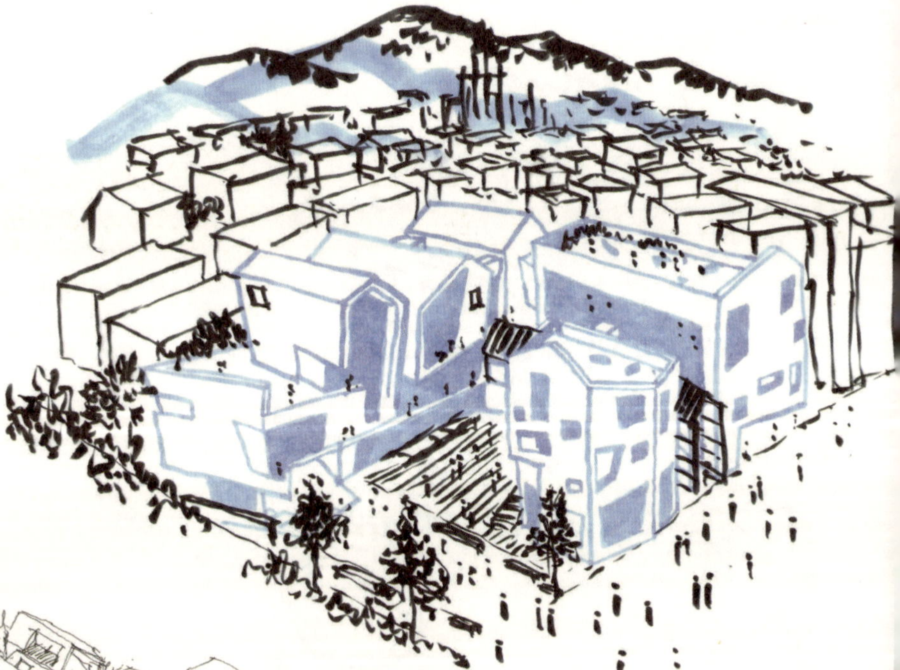

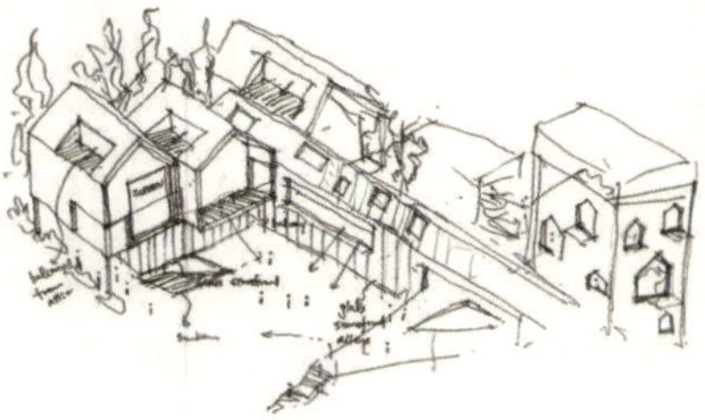

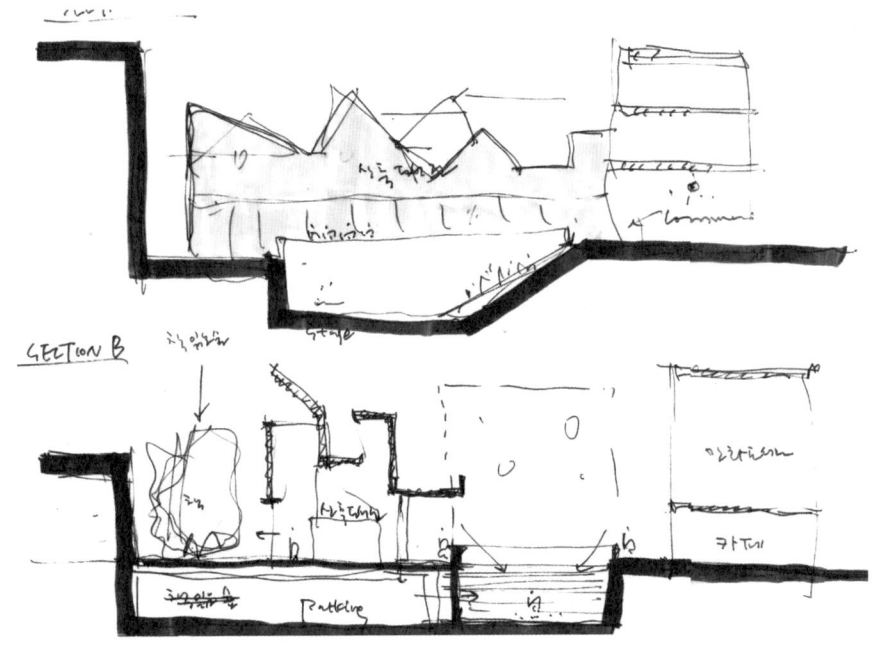

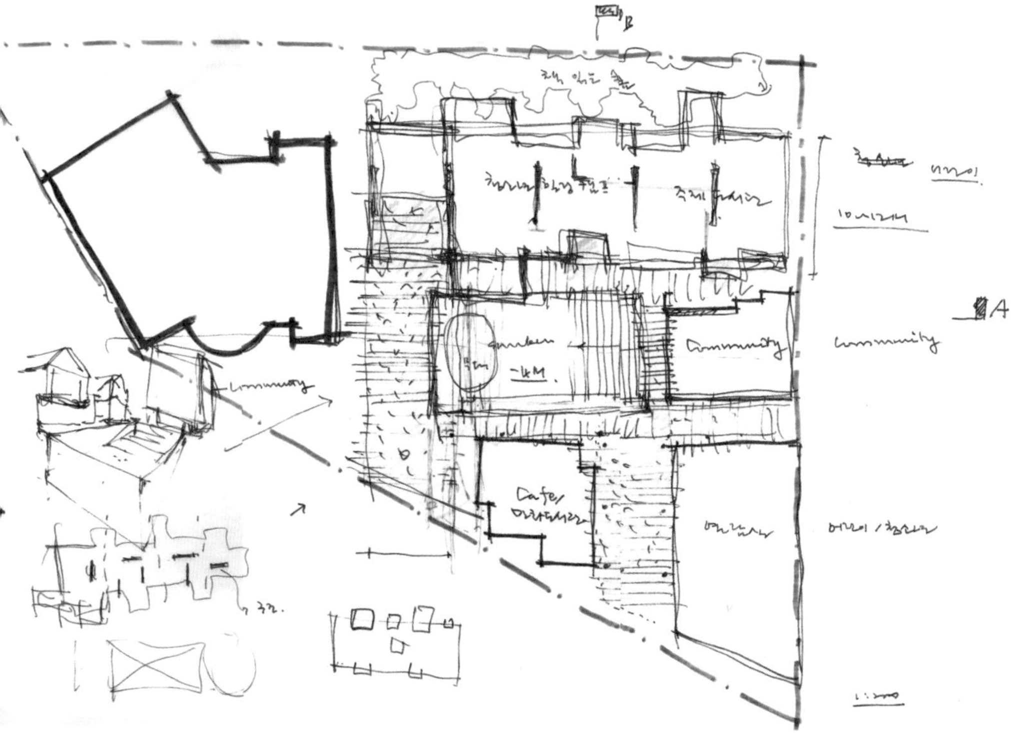

배치계획.
"책 읽는 골목"

느티나무
대나무
+ 마삭줄.
평철난간
대나무
+마삭줄
바닥재
A P.C. Paving.
B 마사토.
느티나무
주민자치
텃밭
시설물
① 자전거 정류장
② 유모차 대기소
③ 벤치.

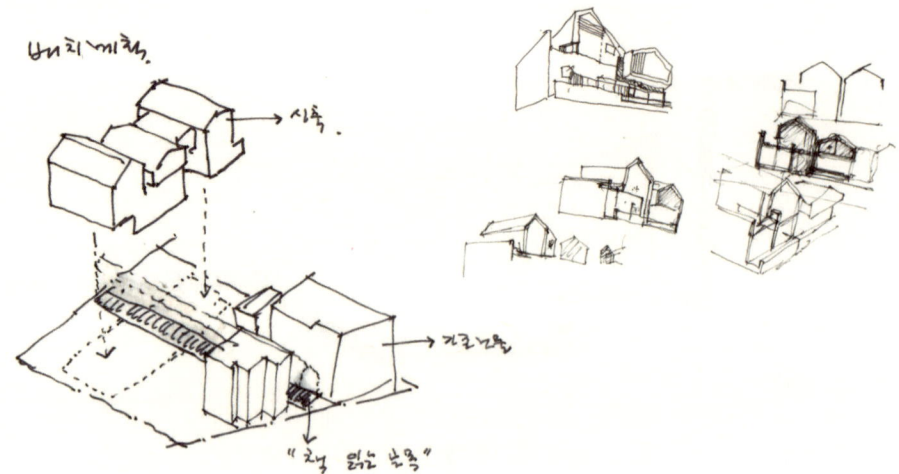

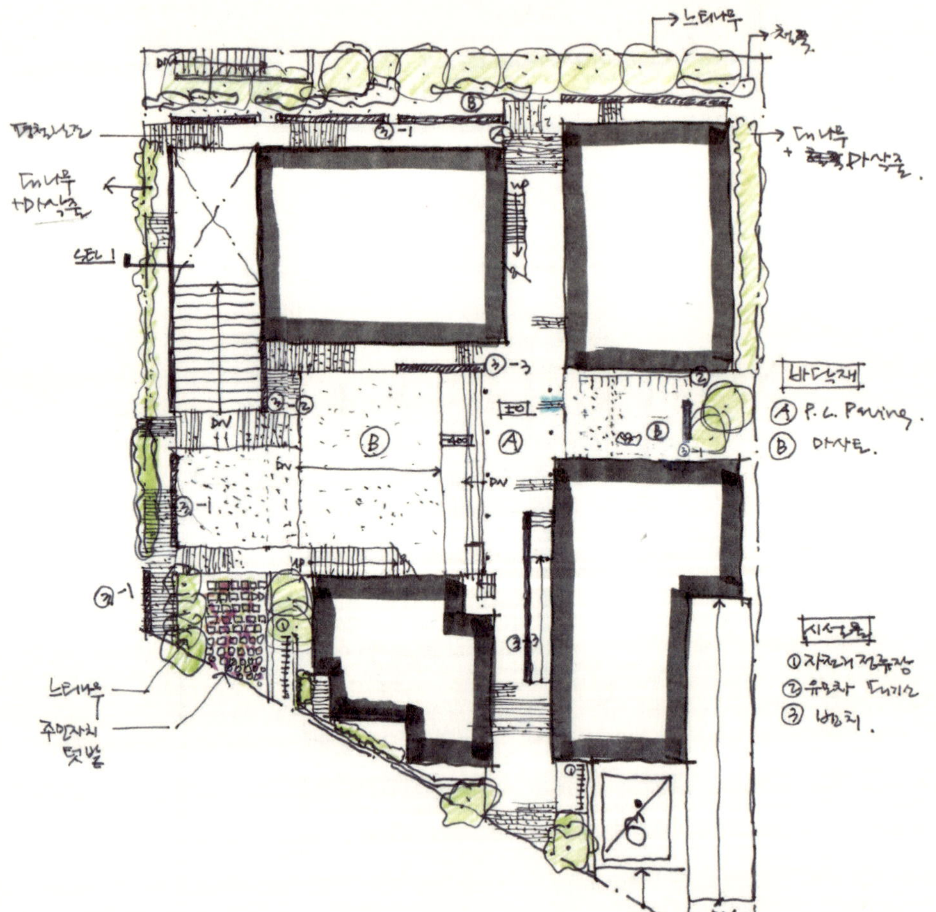

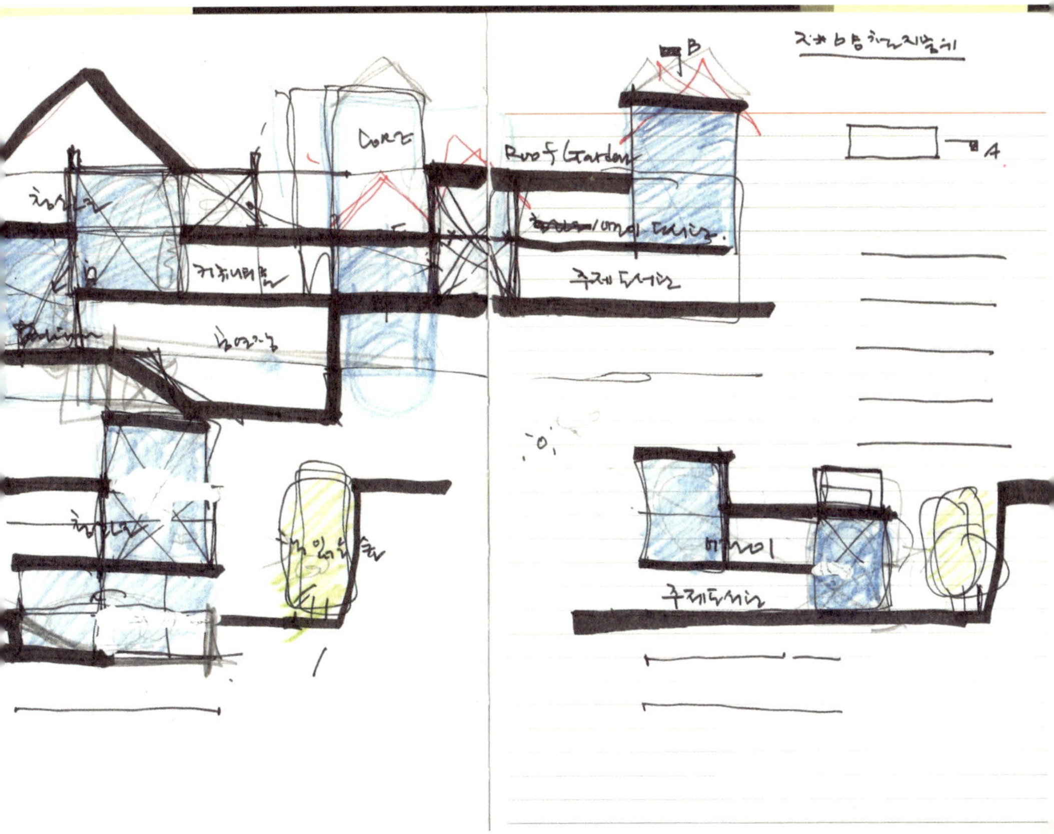
Roof Garden
주제도서관

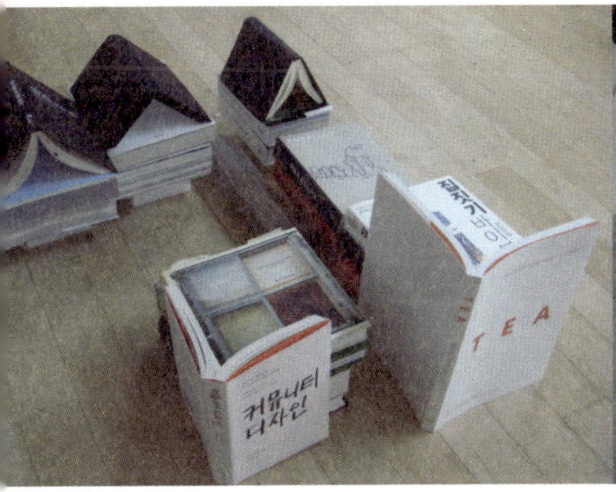
커뮤니티 디자인
TEA

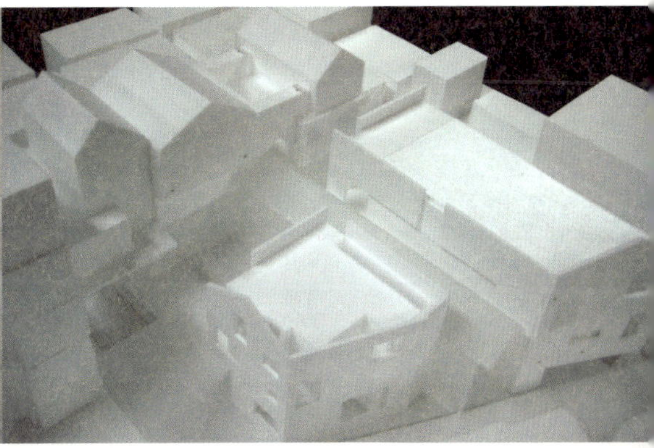

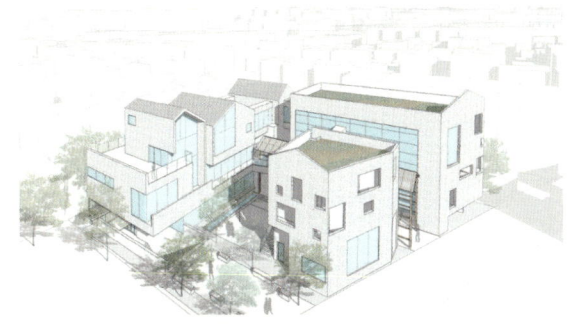

BOARD

www.b-o-a-r-d.nl

Bernd Upmeyer is the founder and principal of the Rotterdam-based Bureau of Architecture, Research, and Design (BOARD). He is also the editor-in-chief and founder of MONU Magazine on Urbanism. He studied architecture and urban design at the University of Kassel (Germany) and the Technical University of Delft (Netherlands). Since June 2012 Upmeyer and his office BOARD are part of the group, led by STAR - strategies + architecture, that has been chosen as one of the new six teams of architects and urban planners appointed by the Atelier International Grand Paris (AIGP) to be part of the Scientific Committee for the mission: Grand Paris: pour une métropole durable.

BOARD (Bureau of Architecture, Research, and Design) was founded in 2005 and is active in many fields: as an architecture and urban design practice, as a research board and as a platform for comparative analysis on urban issues through its bi-annual journal MONU – Magazine on Urbanism. BOARD won several prizes recently in prestigious international architecture and urban design competitions. Among others: The House of Arts and Culture in Beirut, Lebanon; The Estonian Academy of Arts in Tallinn, Estonia; and The New Headquarter for Wexford County Council in Ireland. In 2008 BOARD was listed as one of the "Top Ten emerging offices under 40 in the Netherlands by NIB (New Italian Blood) in collaboration with the A10 Magazine.

What is a notebook to you?

A notebook actually means very little to me. **I am not using one.** The only time I ever used a notebook was during my first year of studying architecture, when my teacher forced us to use one. But I abandoned it very quickly, because I did not believe in its value. In fact, I considered it counterproductive and something that blocks transparency and the exchange of ideas during design processes within teams. In notebooks information becomes very secret, hidden and private. When I hear sentences such as "Architects use notebooks as their treasure chests of ideas", I think of architects of the 20th century such as Louis Kahn, who was known for making numerous trips to Europe during his career in the United States, sketching and writing down everything he saw in a sketch- or notebook, probably because **he did not have pocket-sized cameras or smartphones yet**. That is why notebooks appear to me today as something outdated and old-fashioned, something like a Friendship book in a time when everybody has a profile on Facebook, something like using a sketch as an architectural representation when we have renderings.

Any episodes or memories related to a notebook?

As I am never using a notebook and have rarely used one in the past, I cannot divulge any episodes or memories related to it here. Nevertheless, I use sketching but not in a notebook, but mostly on a sketch roll. **That makes the sketches generally rather temporary and of an ephemeral nature**, as they are usually thrown away directly after they fulfilled their purpose, which is mainly to get a step further in a design, research, or study process. **It does not matter if they are rather ugly**, as long as they communicate the right thing. But they don't have to be representational anymore, meaning that they are not much used for final presentations, as they were in the past, before the time of CAD, Adobe, and rendering software. Sketching can become much more enjoyable than it was in the past as a lot of pressure has been removed from it. Today, it becomes ever more clear that an important effect the integration of the computer has had on the architectural design process has less to do with form, organic shapes and complex geometries, as once was hoped, but with the liberation of the sketch from the notebook, where it was damned to permanence and burdened to represent.

I am not using one.

It does not matter if they are rather ugly.

당신에게 수첩이란 무엇인가?

나에게 노트북은 별 의미가 없어서 **사용하지 않는다**. 내가 유일하게 스케치북을 사용했던 적은 대학교 1학년 때, 교수님이 억지로 사용하도록 했다. 하지만 나는 그 가치를 믿지 않아 얼마 가지 않아 사용하지 않기 시작했다. 나는 오히려 방해 된다고 생각했다. 특히 여러 사람이 함께 디자인하고 아이디어를 나누는 과정에서 투명하지 못하고 역효과를 낳는다고 생각했기 때문이다. 수첩 안에 적히는 정보들은 매우 비밀스럽고 숨겨져 있고 개인적이게 된다. "건축가들은 자신의 수첩을 아이디어 보물 상자라고 생각한다" 라는 말을 들을 때면, 나는 루이스 칸과 같은 20세기 건축가들이 생각난다. 루이스 칸은 미국에서 활동할 때에 유럽 여행을 자주 가며 그가 보고 느낀 모든 것들을 수첩에 그리고 쓴 것으로 유명하다. 하지만 나는 **그가 그렇게 수첩을 사용한 이유는 작은 카메라나 스마트폰이 없었기 때문이라고 생각한다**. 그래서 나에게 스케치북은 오래되고 유행이 지난 것처럼 느껴진다. 마치 모두가 페이스북 프로파일을 가지고 있을 때, 우정의 책을 가지고 있는 것이라던가, 건축 디자인을 보여주기 위해 렌더링을 할 수 있을 때 스케치를 그리는 것과 같다고 생각한다.

수첩에 관련된 에피소드가 있다면 들려달라.

나는 수첩을 거의 사용 하지 않기 때문에 수첩과 관련된 에피소드나 추억은 생각나지 않는다. 하지만 스케치를 아예 하지 않는 것은 아니다. 그저 수첩이 아닌 트레이싱지에 할 뿐이다. **그렇게 그려진 스케치들은 그 순간에만 사용되는 경우가 많다.** 주로 디자인 발전 과정에서 사용되어 그 역할이 끝난 이후에는 버려진다. **못 그려도 상관없다.** 전하고자 하는 뜻만 잘 전달되면 되는 것이다. 그리고 최종 프레젠테이션 때 사용되는 이미지처럼 모든 것을 표현하지 않아도 좋다. 이제는 캐드, 어도비, 그리고 다양한 렌더링 소프트웨어가 있기 때문이다. 전처럼 완벽하게 그려야 한다는 부담감이 사라졌기 때문에 좀 더 즐기면서 스케치를 할 수 있다. 오늘날 컴퓨터가 건축 디자인 과정에 있어 유기적인 형태나 복잡한 기하학을 형성하는 데 도움을 주는 것보다 스케치를 수첩으로부터 자유롭게 해주는 중요한 역할을 하고 있다는 것이 점점 더 명확해지고 있다. 영구적인 것으로부터 벗어나고 보여줘야 하는 것에 대한 부담감을 덜었다.

© BOARD

When and where do you use your notebook the most?

Ever since the sketch was liberated from the straightjacket of the sketchbook or the architectural notebook, it could be produced everywhere at any time. If I am on a plane or train, the corners of a newspaper are, for example, great places to write or sketch something that you wish to remember. However, I do most sketching in my office during the day. This activity is mostly driven, apart from the need to communicate ideas to others, by the fear of losing thoughts and ideas. **But that kind of sketch shares the same destiny as the one on a sketch roll: once it has fulfilled its purpose and was translated into a digital architectural drawing, diagram, or image, or communicated to another person, it is thrown away. Nevertheless, I don't think we should mourn the short-lived nature of contemporary sketches;** just as we do not mourn the short lives of mayflies that only live from a few minutes to a few days, depending on the species. Adult male mayflies have two penises and during the few days they live in spring or autumn they are everywhere, dancing around each other and copulating in large groups on every available surface.

© Sonia Arrepia

수첩을 가장 많이 사용하는 공간과 때는?

스케치가 스케치북이나 건축 수첩으로부터 자유로워진 순간부터 언제 어디서나 그려질 수 있는 것이 스케치다. 비행기나 기차에 타고 있을 때, 예를 들면 신문지 모퉁이가 기억하고 싶은 것을 그리거나 쓰기에는 정말 좋은 곳이다. 하지만 나는 대부분 낮에 사무실에 있을 때 스케치를 한다. 사람들과 소통해야 한다는 이유 말고는 떠오른 생각이나 아이디어를 잊지 않기 위해서 스케치를 하는 것이 대부분이다. **하지만 이러한 스케치들 또한 트레이싱지에 그려진 것들처럼 자신의 역할을 다하고 컴퓨터 도면이나 다이어그램, 또는 이미지로 옮겨졌을 때에는 버려진다. 그렇다고 이러한 스케치들을 불쌍하게 생각할 필요는 없다.** 우리가 몇 분에서 며칠밖에 살지 못하는 하루살이들을 불쌍하게 생각하지 않는 것처럼 말이다. 수컷 하루살이들은 두 개의 생식기를 가지고 있어 자신이 살 수 있는 며칠동안 가능한 모든 곳에서 교미한다.

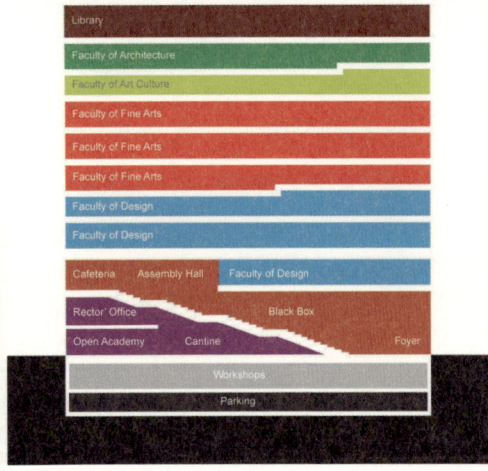

© BOARD

What influence does a notebook have in your projects and life as an architect?

However ephemeral and "ugly", the sketch itself has still a very strong influence on my projects and therefore also on my life, and I believe the same applies to a lot of other architects of my generation; by contrast, the sketchbook has very little influence. **I think it is no coincidence that the moment in the mid-1980s when computer-aided design programs appeared, the last notebook manufacturer, supposedly one of the original "Moleskine" producers in France, stopped production.** It took a lot of years before an Italian company brought that kind of sketchbook back on the market, establishing it as a trademark, breathing new life into a dead product, and marketing it as something legendary that famous avant-garde artists and writers used in the past. Recent impressive growth numbers have proven its commercial success, which, I believe, is the reason I am now writing about sketchbooks to begin with. I think the sketchbook is still essentially dead, but enjoys a boom in sales, because of brilliant marketing and branding to a nostalgic mass of people that wishes to look as creative as the avant-garde that was supposed to have used it in the past. Perhaps this reflects a certain desire for something permanent and solid in an increasingly ephemeral temporary digital world in which one click can destroy everything within a split second.

수첩은 당신의 작품과 삶에 어떤 영향을 끼치는가?

얼마나 일시적이고 '못생긴' 스케치일지라 하더라도, 프로젝트에 매우 많은 영향을 끼치기 때문에 나의 삶에도 영향을 끼치고 있다고 생각한다. 우리 시대 대부분의 건축가들은 같은 생각을 할 것이라 믿는다. 반면에 스케치북 그 자체에는 거의 영향력이 없다. **1980년대에 캐드 프로그램이 처음 나왔을 때, 오리지널 스케치북 중 하나인 프랑스의 '몰스킨'이 생산을 중단한 것이 나는 그저 우연이라고 생각하지 않는다.** 이탈리아 회사가 시장에 다시 내놓기까지에는 많은 시간이 걸렸다. 한때, 아방가르드 했던 예술가들과 작가들이 사용했던 전설적인 제품으로 마케팅 하면서 잊혀진 것을 다시 살려냈다. 최근 놀라울 정도로 늘어나고 있는 것을 보면 '몰스킨'이 성공한 것을 보고 있다. 내가 지금 스케치북에 대해 쓰고 있는 것도 이러한 이유 때문이 아닐까 생각한다. 그래도 나는 아직도 스케치북의 가치는 떨어졌다고 생각한다. 하지만 판매가 늘어나는 이유는 전에 아방가르드 했던 사람들처럼 창의적으로 보이고자 하는 사람들의 마음을 사로잡는 브랜딩과 마케팅을 통하기 때문이다. 어쩌면 이러한 모습이 한순간에 모든 것을 지워버릴 수 있는 일시적인 디지털 세상 속에서, 사람들의 **무언가 영구적으로 남기고 싶어하는 마음**을 보여주는 것일 지도 모른다.

Are there anything else other than a notebook that you use to keep a record of your thoughts and ideas?

In addition to the abovementioned corners of a newspaper and sketch rolls: used sheets of paper and paper from printing errors are also great media to sketch and write on. I do that a lot. In my office we once had, for example, an A3 printer and a lot of A3-sized paper. But ever since the printer broke down, the sheets of A3 paper remain unused in the storage, piling up without being used. I recently counted around twenty boxes of five hundred A3 sheets of paper each. So what we do is, we cut those sheets into handy A4s and use them as sketch paper, and to keep a record of thoughts and ideas as well. But then again, those papers are usually thrown away after the sketches and thoughts are translated onto another medium. Only a handful are kept in a project folder. All this shows how in my office the architectural sketch is demystified and used merely as a tool to remember and communicate things – something utterly temporary that does not require prestigious and glamorous treatment.

수첩 외에 자신의 생각을 기록하는 방법과 도구는 무엇이 있는가?

앞서 말한 신문지 모퉁이나 트레이싱지 외에 일반 종이나 이면지 또한 스케치하거나 적기 좋은 곳이다. 나는 정말 많이 사용한다. 예를 들자면 우리 사무실에 전에 A3 프린터와 용지가 많이 있었다. 하지만 프린터가 고장이 난 후로 A3 용지는 사용되지 않은 채 그저 쌓여져 있기만 했었다. 500장 짜리 A3 용지 묶음이 한 25박스 정도 있었다. 그래서 우리는 이 용지들을 사용하기 좋은 A4 용지로 잘라 스케치 페이퍼로 사용하여 아이디어들과 생각들을 정리하기 좋게 만들었다. 하지만 이 스케치들도 다른 곳으로 옮겨진 후에는 버려지는 일시적인 것들이었다. 어떠한 것들만 프로젝트 폴더 안에 넣어 보관한다. 우리 사무실에서는 스케치들을 기억하고 소통하기 위한 도구들로만 사용하는 것을 볼 수 있다. 화려한 터치 없이 매우 일시적인 것이 바로 스케치다.

b4 architects

www.b4architects.com

The group was founded in Rome in 2003, from 2008 the active partners are Gianluca Evels and Stefania Papitto. In 2012 the office becomes b4architects associated. Measuring us from urban and landscape design until restoration, interior design and object design, in a process that involve specific competences, always respecting the environmental characteristics. We are interested in producing works that contribute to the debate of the complexity of modern life. Our approach to any project is to involve all parties in a creative collaboration to define the objectives of the project with a balanced combination within critical readings of the local context and the "outsider" perspective of us. The work on pre-existent spaces and the interior design projects try to explain all the available elements in a new synthesis: the traces of the history of the building, the expectations of the client, psycho-sensorial aspects of the architecture united in a continuous spatial and visual tale. We also dedicate to further activities, like some different experiences at the university of Rome or taking part at some international workshops, with the aim to be active in the debate about contemporary architecture. The office attend to building energy consulting both for new construction and for renewal of existing building. Arch. Papitto is certified as ClimateHouse-Expert-Planner for ClimateHouse Agency in Bozen.

What is a notebook to you?

A notebook is a safe-conduct for the mind and for ideas. The need is to take it everywhere, at any time you may need to use it to fix ideas, concepts, thoughts, visions. It's still very versatile, more than any electronic tool available. And it is immediate, do not need of battery or energy to work, only your active brain. It should be light and easy to handle, ready to use. It can frame design thoughts or visions of travel. In the case of use in travel, it records visually what it is important for you, but through a filter that selects only what really matters, even just suggestions. Over time it becomes a repository of memory, also useful to recall ideas to reconsider even after a long time. Normally the paper is white, of a robust weight without signs or lines to allow maximum freedom of expression. The drawings or words are laid out with a pen usually thin and smooth, they are sometimes also used colors to emphasize certain concepts. The size is normally small and pocket, but often are also used common A4 sheets held together by a provisional clamp on a rigid support to use it everywhere.

당신에게 수첩이란 무엇인가?

수첩은 생각과 아이디어들을 위한 안전 통행권과도 같다. 어디든지 가지고 다녀야 하고, 아이디어나 컨셉과 생각 그리고 비전들을 언제 바꿔야 할지 모른다. 그 어떠한 전자기기들보다 다용도로 사용할 수 있고, 활발한 두뇌만 있다면 배터리나 에너지 없이 바로 사용할 수 있다. 매우 가볍고 쉽게 사용할 수 있어야 한다. 디자인 생각이나 여행하면서 상상했던 것들을 담아줄 수 있다. 여행하면서 사용 할 때에는, 자신에게 가장 중요한 것을 시각적으로 기록하지만, 여기있는 정말 중요한 것들만 걸러진다. 시간이 지나면서는 기억을 보관하는 곳이 되어, 오랜 시간이 지난 후에도 아이디어들을 다시 기억해 낼 수 있도록 해준다. 종이는 주로 흰색이고, 아무런 표시나 줄이 없는 두꺼운 종이로 만들어져 표현의 자유를 최대화 시켜준다. 그림이나 단어들은 펜을 사용하여 주로 얇고 부드럽게 종이에 담겨진다. 특정한 컨셉을 강조하기 위해 색을 사용 할 때도 있다. 주로 작고 주머니 크기만한 스케치북을 사용하지만, A4용지들을 받침대와 함께 클립으로 고정시켜 사용할 때도 많다.

Any episodes or memories related to a notebook?

Over the years they have been preserved with chronological order all the notebooks in various dimensions. It happened that one of them has been lost in the classrooms of the Faculty of Architecture in Rome a few years ago. It contained all the sketches of a journey between Berlin and Canada, with important experiences recorded in the drawings and in the notes. The sense of to be out of element for the loss was great and not yet fully ridden out: part of the 'archive' of the mind is lost. The feelings, the ideas recorded on the architecture visited in Berlin (Mies, Libeskind, Eisenmann, Scharoun, Piano, etc..) and Canada have been and still are an important cultural baggage for the project activity. From that event periodically **all the sketches are scanned and stored digitally**, to avoid further unpleasant similar events.

수첩에 관련된 에피소드가 있다면 들려달라.

지난 몇 년동안 사용한 다양한 크기의 모든 수첩들은 시간 순서대로 정리되어 있다. 그 중 하나를 몇 년 전에 로마에 있는 건축대학 교실에서 잃어버린 적이 있었다. 베를린과 캐나다 여행에서 경험했던 중요한 사건들의 스케치들과 노트들이 담겨있는 수첩이었다. 소중한 것을 잃었다는 슬픔이 컸다: '기록'하는 마음을 잃었다. 베를린과 캐나다에서 방문했던 건축들 (미스, 리베스킨트, 아이젠먼, 샤룬, 피아노 등)을 보면서 느꼈던 감정들과 아이디어들은 프로젝트 진행함 하는데 있어 매우 중요한 문화적 자산이었다. 비슷한 일이 다시는 일어나지 않도록, 이 사건 이후로 **모든 스케치들을 정기적으로 스캔하여 디지털 파일로 보관한다.**

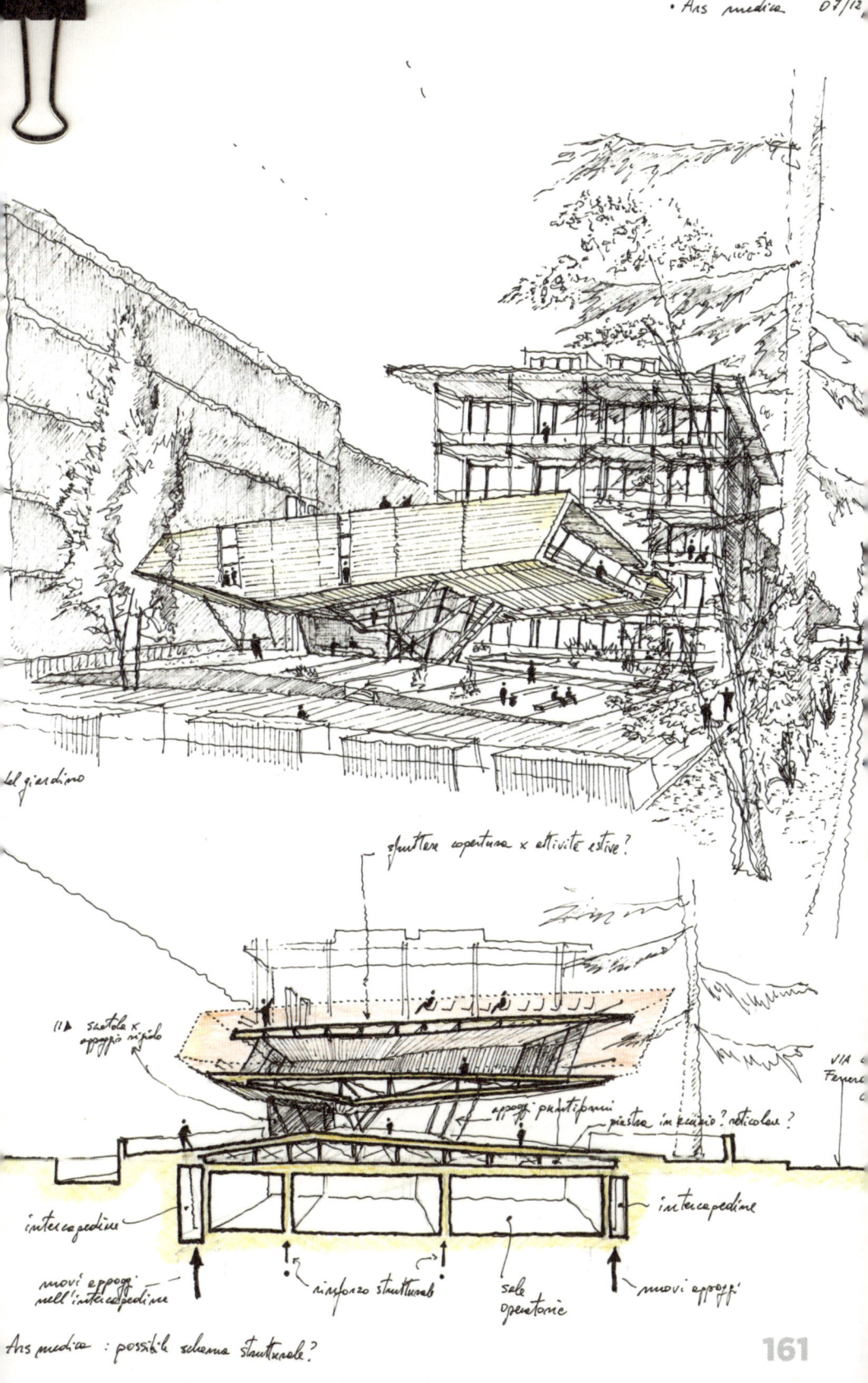

*barcellona

When and where do you use your notebook the most?

The beginning of each new project normally involves a phase of brainstorming that occurs to subsequent sketches on paper until you come to a concept that is checked as soon as possible to the PC. Subsequently, the sketches are a tool that continues to take place alongside digital design. The places where the notebook is used are always different. It happens that the most informal places are more fruitful for design ideas, particularly in the more intuitive phase: a café bar, the stairs in a public place, park, traveling on a train or even on holiday. The office or more 'official' places of the work are used for the process in depth analysis of ideas.

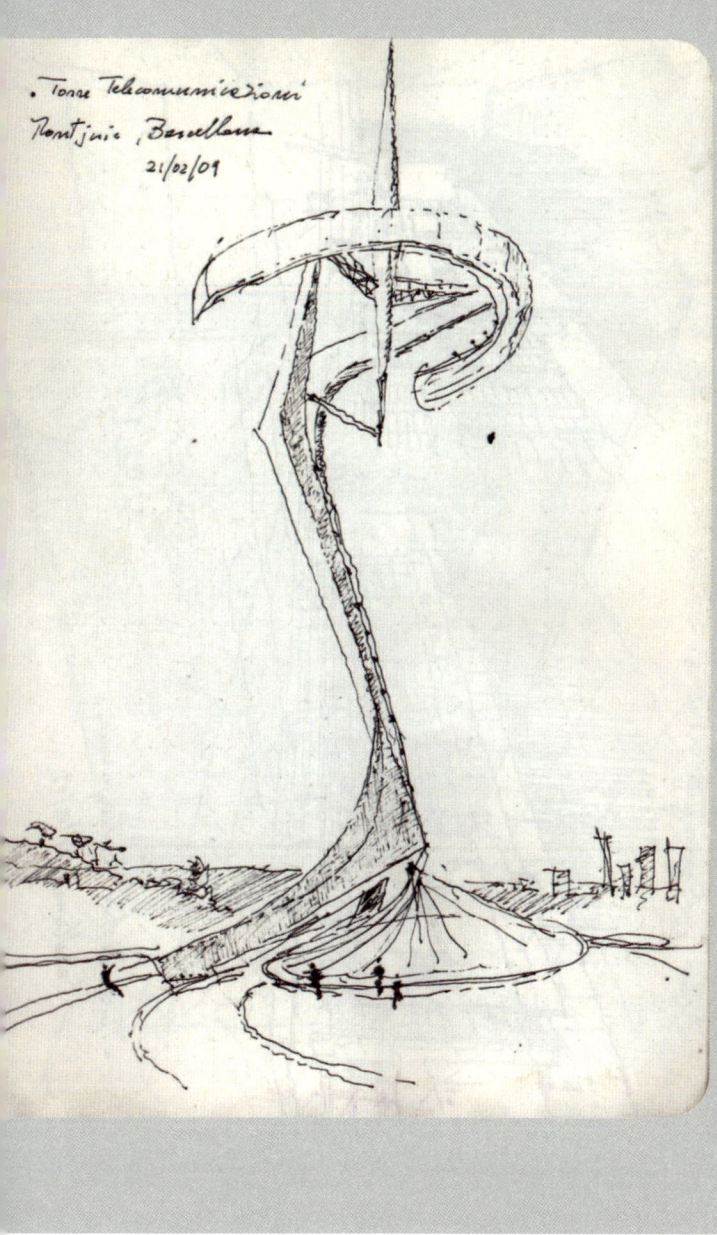

수첩을 가장 많이 사용하는 공간과 때는?

새로운 프로젝트의 시작은 주로 브레인스토밍으로 시작된다. 컴퓨터로 확인해볼 수 있는 **컨셉이 나올 때까지 스케치를 하는 것이다.** 하지만 디지털 디자인을 하면서도 스케치는 계속 한다. 스케치북을 사용하는 공간은 매번 다르다. 디자인 아이디어들이 가장 많이 생각나는 곳들은 주로 편안한 분위기의 공간들이다. 카페나 계단, 공원이나 기차 안 같은 공간들이 초기 단계에서는 특히 더 좋다. 사무실이나 좀 더 '공식적인' 작업 공간들은 아이디어를 깊게 연구하고 발전 시키는 과정에 좋다.

*corinth

What influence does a notebook have in your projects and life as an architect?

In the evolution of the office's work, sketch have registered an important role for the development of the project. In life as an architect it becomes the privileged way to observe and communicate, a tool that becomes an inseparable part of themselves, whether it be a way to observe and record external things or inner visions. It becomes **almost an extension of the body** and for this reason one can not do without and you bring it everywhere in daily life. It 'a tool to get to know spaces and materialize ideas.

수첩은 당신의 작품과 삶에 어떤 영향을 끼치는가?

사무실 작업들을 보면, 스케치는 프로젝트를 발전시키는데 중요한 역할을 하게 되었다. 건축가로 살아가면서 스케치북은 특권이다. 외부적인 것이든 내면의 모습이든, 모든 것을 관찰하고 소통하고 기록하는 도구다. **몸의 일부분**이라고 해도 될 만큼 가까운 것이기 때문에 어디를 가던지 항상 가지고 다녀야 한다. 공간을 알게되고 아이디어를 구체화 시키기 위한 도구다.

Are there anything else other than a notebook that you use to keep a record of your thoughts and ideas?

Also the camera is a powerful tool to record feelings and visions store. And it is immediate, versatile and it records a substantial amount of useful information. But photography in the hands of an architect has less 'selective' chance of the topics that interest in confront of the possibilities of the drawing, and a most limiting expressive possibility for the ideas to communicate. Of course it is a very personal and expedient opinion of course.

All the sketches are scanned and stored digitally.

수첩 외에 자신의 생각을 기록하는 방법과 도구는 무엇이 있는가?

카메라 또한 감정이나 모습들을 담기에 매우 강력한 도구다. 또한, 바로 사용할 수 있고, 다용도이며 상당한 양의 필요한 정보들을 기록할 수 있다. 하지만 사진은 건축가에게 있어 제한적이다. 그림을 그리는 것과는 달리, 주제들을 부분적으로만 '선택' 하여 기록할 수 없기 때문이다. 아이디어를 표현하여 소통할 때 제한적이기도 하다. 물론 이 모든 것은 나의 주관적이고 편의주의적인 생각이다.

*delfi

*katsura, Japan

*delfi

*Milano

*Rotterdam

*Venezia

*Porto

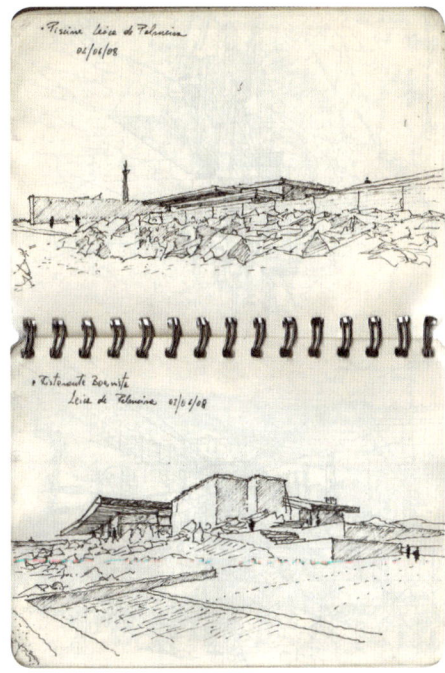

*Rotterdam

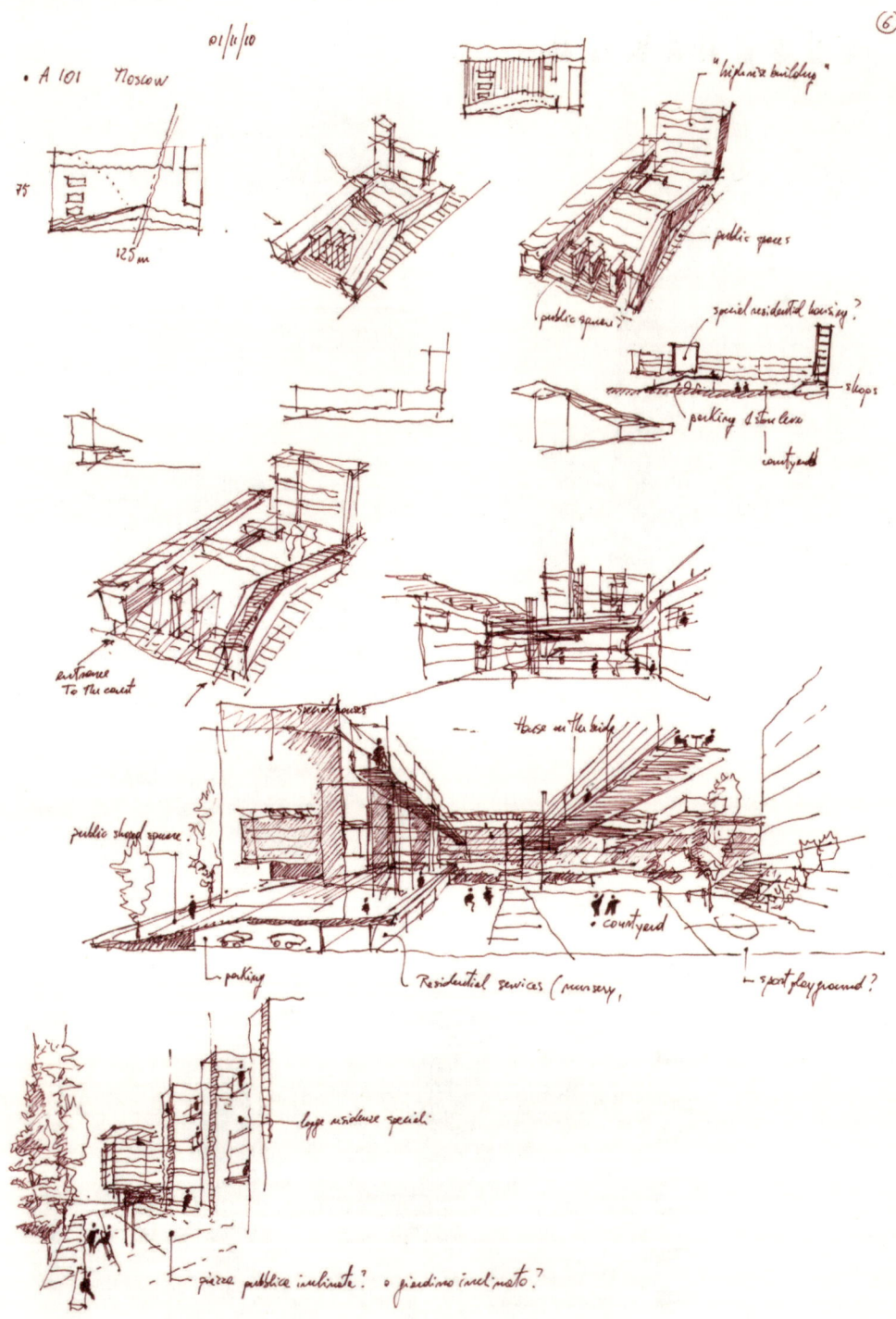

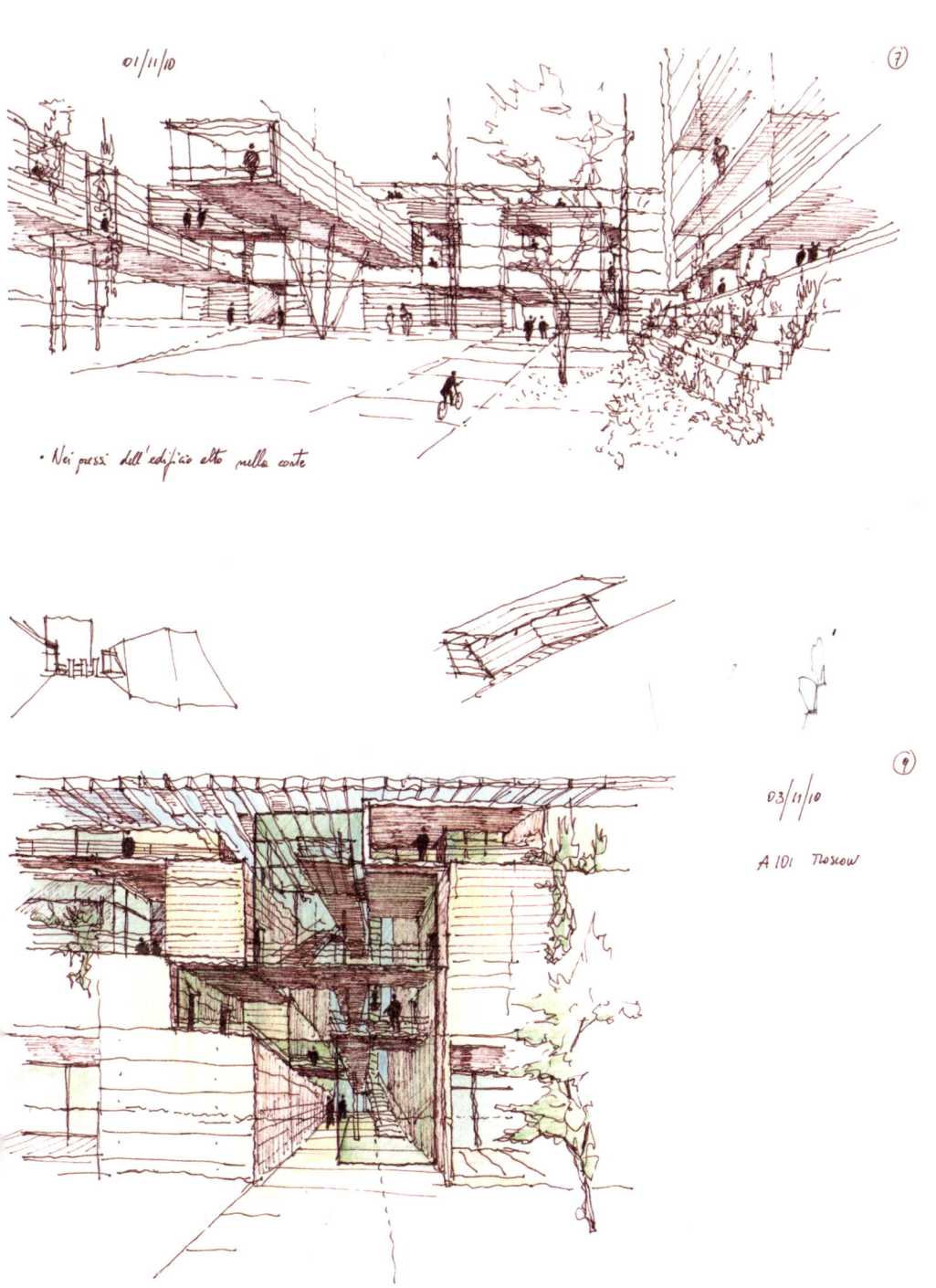

- Nei pressi dell'edificio alto nella corte

- Uno degli atri condominiali → verificare eventuale chiusura retrostante scale condominiale

Cheong-na

① Cheongra City Tower, Incheon Korea

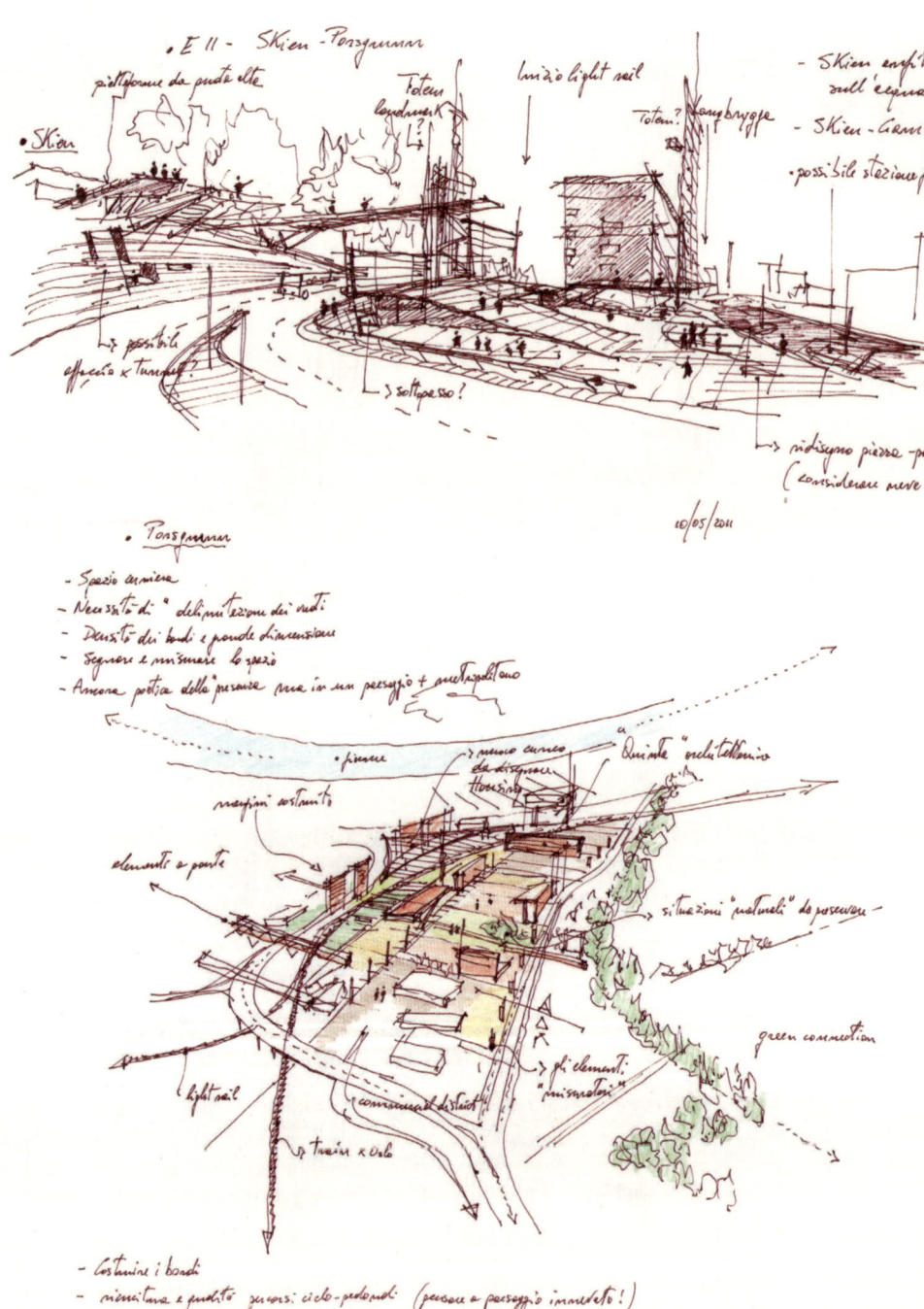

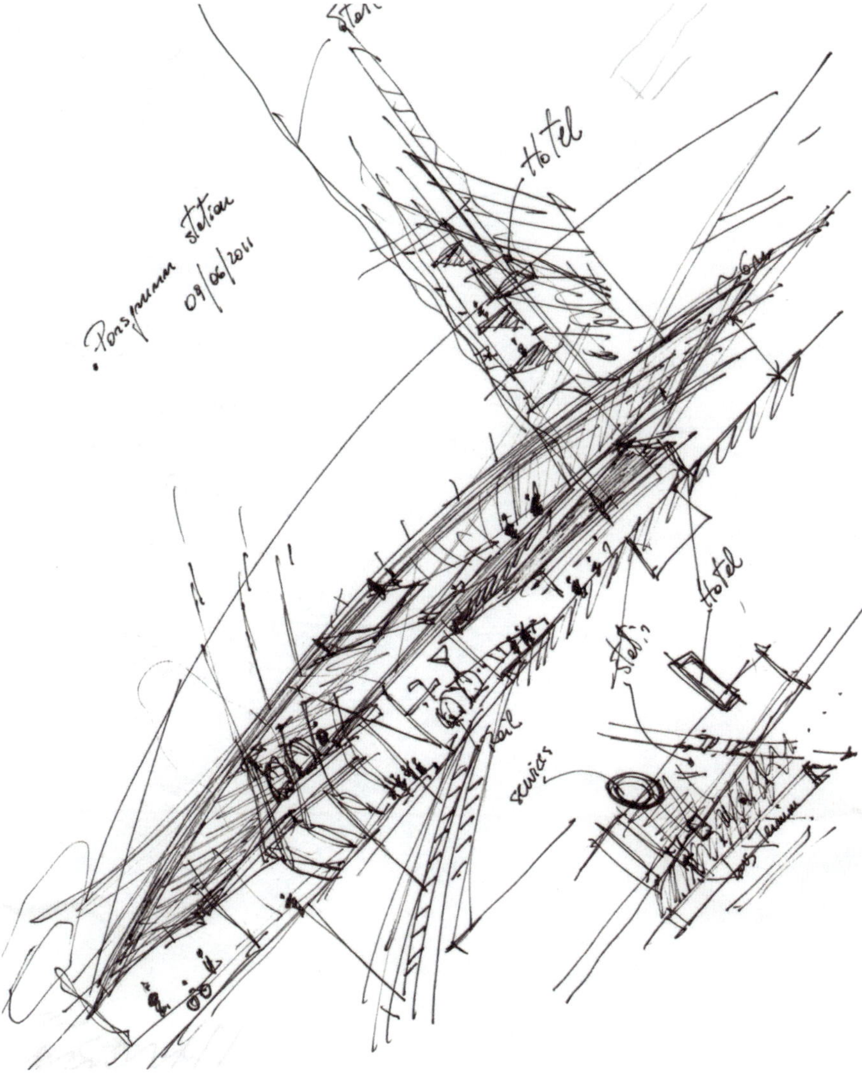

*Outside the box

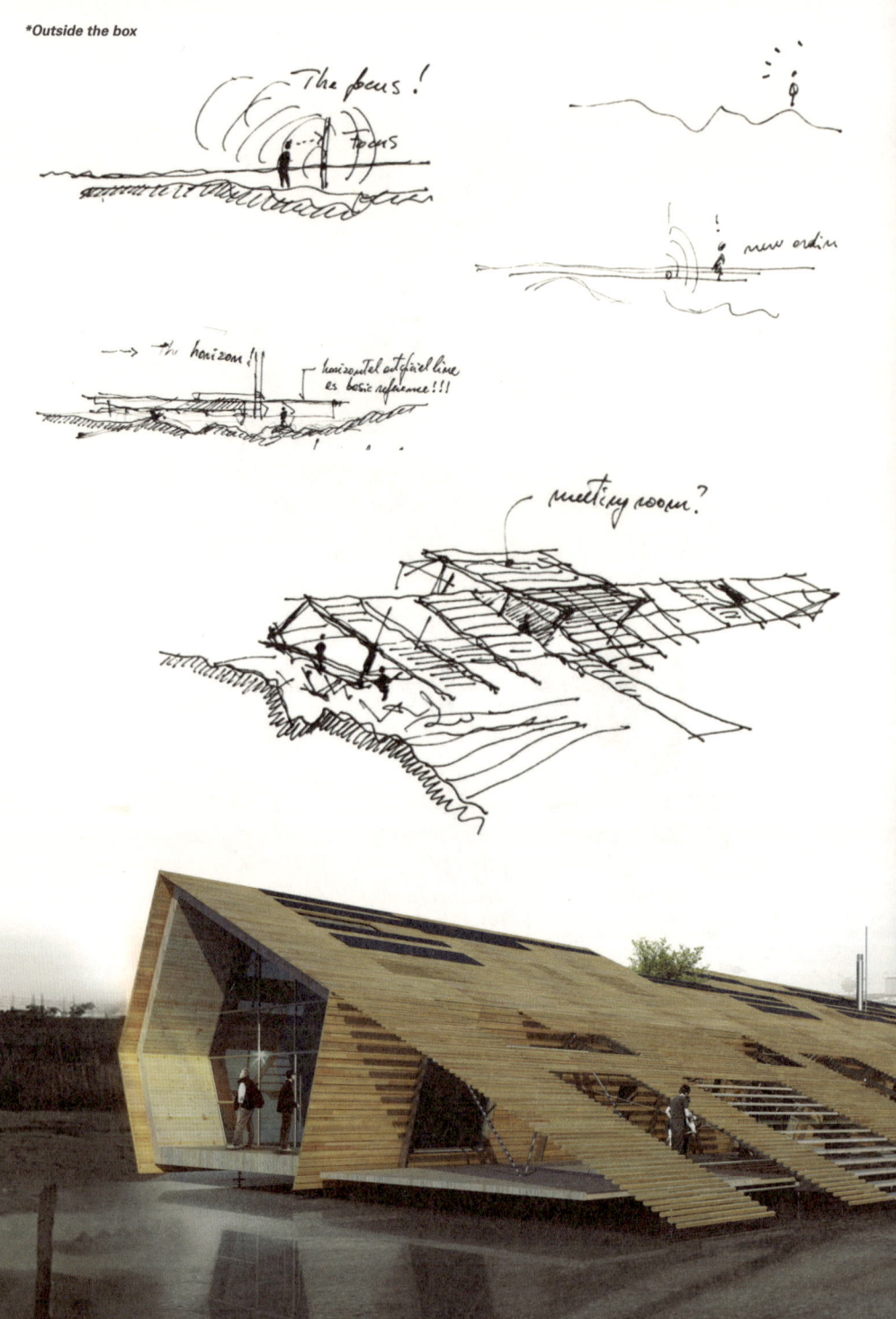

*Terracina

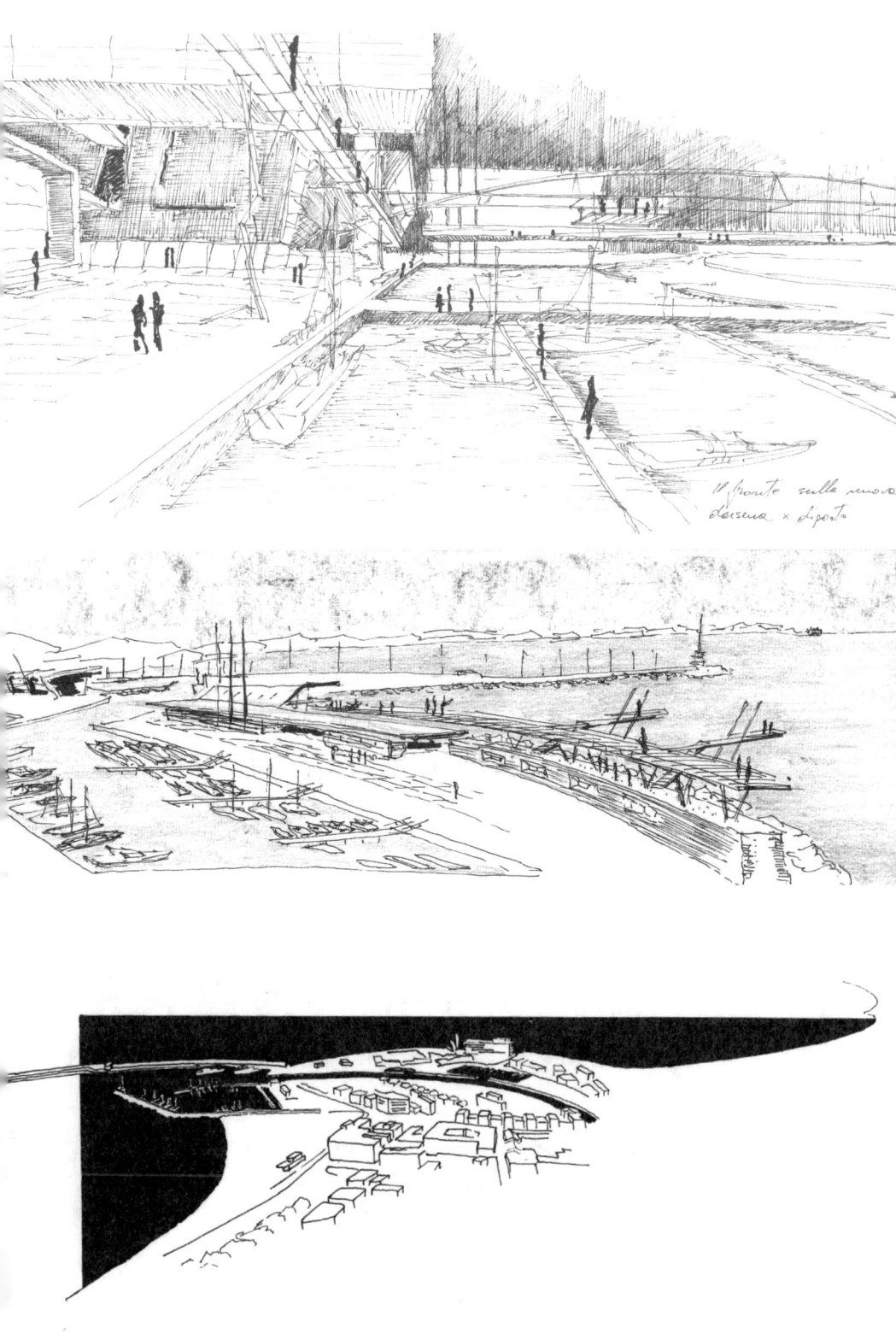

*Vancouver

Heather Woofter [Axi:Ome]

www.axi-ome.net

Heather Woofter studied biochemistry as a teenager at the University of Maryland and chemical engineering at Virginia Polytechnic Institute. She received a Bachelor of Architecture from Virginia Polytechnic Institute and Masters of Architecture from Harvard University Graduate School of Design. She was a project architect and manager for Bohlin Cywinski Jackson in Wilkes-Barre, Pennsylvania, Marks Barfield in London, UK and Robert Luchetti Associates in Cambridge, Massachusetts.

She is a Registered Architect in Commonwoalth of Pennsylvania and Missouri and has Royal Institute of British Architects Parts I and II. Heather taught at the Harvard Graduate School of Design Career Discovery Program, Boston Architectural College and Roger Williams University. She was an Assistant Professor at Virginia Polytechnic Institute, Visiting Professor at Aristotle University of Thessaloniki in Greece and Konkuk University in Seoul, Korea. Currently, she is a tenured Associate Professor of Architecture and a Chair of the Graduate School of Architecture at Washington University in St. Louis. She is a founding director and owner of Axi:Ome llc of St. Louis with Sung Ho Kim since 2003.

The notebook is a place to visually remember a moment.

When and where do you use your notebook the most?

I keep a notebook with me throughout the day to make notes and think through design strategies. Equally, **the notebook is a place to visually remember a moment.**

당신은 스케치북은 언제 사용하는가?

나는 메모를 하고, 디자인 계획을 생각하기 위해 종일 수첩을 가지고 다닌다. 동시에 수첩은 **한순간을 시각적으로 기억할 수 있도록 해주는 공간이다.**

Any episodes or memories related to a notebook?

Each notebook represents **a different time**. In reviewing drawings of specific places, I am often transported into the past and I remember the intangible qualities of a place in addition to its form.

수첩에 관련된 에피소드가 있다면 들려달라.

각 수첩은 **다른 시간**을 대표한다. 특정한 곳의 스케치를 볼 때에, 나는 주로 과거로 돌아가 그 형태뿐만 아니라 뭐라고 말할 수 없는 특징까지도 기억한다.

What influence does a notebook have in your projects and life as an architect?

The notebook is a place to think. I believe there is a connection between the hand and spatial thinking.

수첩은 당신의 작품과 삶에 어떤 영향을 끼치는가?

수첩은 생각하는 곳이다. 나는 손과 공간적 사고가 서로 연결되어 있다고 생각한다.

I believe there is a connection between the hand and spatial thinking.

Are there anything else other than a notebook that you use to keep a record of your thoughts and ideas?

Photographs help to record, but **the act of drawing creates a kinesthetic memory**, and for me, has a greater possibility of emotive qualities.

수첩 외에 자신의 생각을 기록하는 방법과 도구는 무엇이 있는가?

사진은 기록하지만, **그림을 그리는 행위는 운동기억을 만들어낸다.** 나에게는 이러한 요소가 훨씬 더 감정을 자극한다.

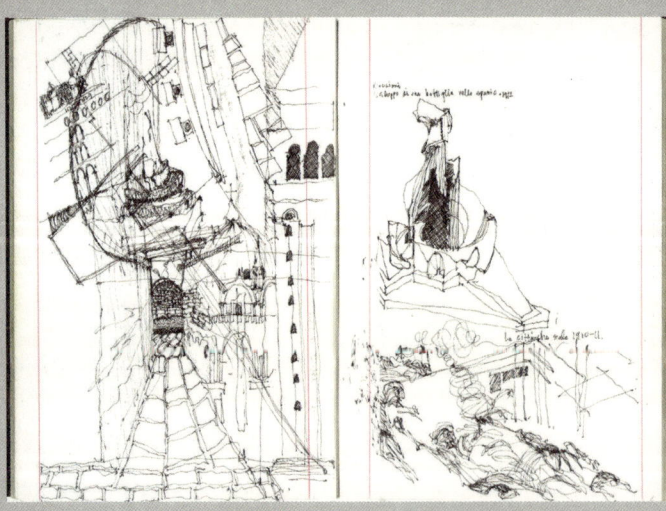
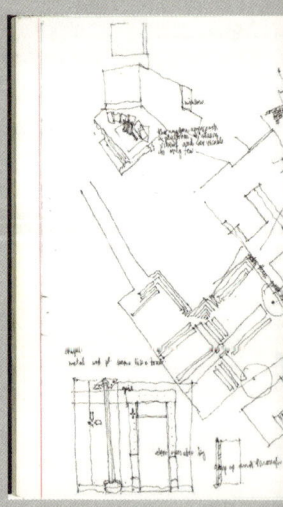

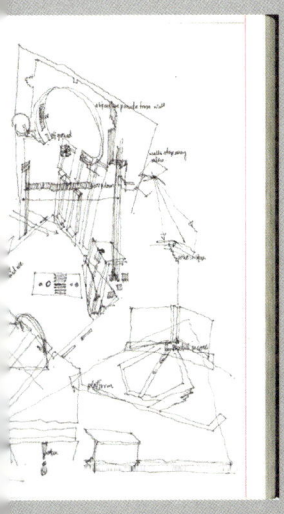

ARPHENOTYPE

www.arphenotype.com / www.horhizon.com / www.digitales-gestalten.de

We have maintained a deep interest in blurring and observing the boundaries between various artistic disciplines. Currently, Arphenotype is focusing upon utilizing techniques and constraints present within the production of synthetic systems and immersive digital virtual environments to influence architectural thoughts and potential modes of production. Since the advent of the computer chip, all of our productive capacities have become more computationally driven, whether through the direct inculcation of codes from our CPUs or through our hands mutating from the psychological affective range of computer thought. We see computation as an inextricable and intrinsic property within our environment that may be used to reconceptualise the current state of architecture. It raises the question of how the body as system is being recomposed. The changes are seen as electronic templates that become a part of the body ? the body is no longer an isolated system; it is being recalibrated. We can apprehend that computers are extended phenotypes of our bodies and our skin. The question of mortality and immortality has been around for thousand of years. The development in techno genetics have profound possibilities not only for reengineering the body but also to reformat it. The new body is in its old skin augmented and perhaps simply redirected. The construction of the body as a series of electronical templates mixed with bones and flesh. The "other body" raises the question of how the body itself is a material ? a Palimsest of electronic scripts and blood ? a series of Phenotypes burning on each other; bending the flesh with digital scratches. The idea of information of systems in the body become part of experiencing causes, this causes the body to be a responsive site. No longer dictated, the body is being electronical formatted.

High seas facilities on the other hand would b
be tied so closely to a port city and which are noxi
residential locations. Oil refining, nuclear fuel rep
water fisheries, and power generation for these plants

Environmenta

What is the potential for despoiling the vast reso
of ocean-based industrial development? Obviously t
the industrial use of chemically active and radioactiv
ecospheres of the terrestrial environment. The detecta
fluorocarbons, the presence of insecticides in fatty ti
other biologic and geophysical horror stories of the las
that the ocean does have a large capacity to assimilate
volumetric size of the hydrosphere, and in part to the re
high-pressure environment at the oceans bottom.

The discovery that chemical processes associated
retarded in the deep sea environment was one of those
full of. In 1968, the research submarine *Alvin* sank in th
The three crewmen escaped, but the ship, with the crew
bottom, a mile below the surface. The vessel was raised a
the apples, and two thermoses of bouillon were in remark
soaked through with sea water. The apples fared no b
refrigerator, but the rest of the edibles were in a far bette

Alvin's sandwiches had spent their ten months under
of 39° F. Testing the role of these conditions on the chem
Woods Hole researchers concluded that these substance
slowly in the deep sea than under standard temperature a

The implications of this for the disposal of waste m
one hand, organic waste d

A place were I note important books, people, phone numbers or even recipes.

What is a notebook to you?

My notebook is something to quickly add notes, hold ideas link comments from lectures and conferences to systematic ideas. **A place were I note important books, people, phone numbers or even recipes.** I have nowadays a notebook function in my smart phone, but it is not the same. Even so, that Arphenotype is at the moment very digital driven, I think the analog way is as important as it always was. The sketchbook enables me to react on the fly - often I am surprised about what I find in my books, when I start looking at them years later.

당신에게 수첩이란 무엇인가?

나의 수첩은 손쉽게 메모를 하거나 아이디어를 담아두는 곳이다. 강의나 컨퍼런스에서 적어둔 것들을 모아 아이디어들을 정리해둔다. **중요한 책이나 사람, 전화번호나 레시피까지도 적어두는 곳이다.** 요즘에는 스마트폰에 수첩을 대신하는 어플이 있긴 하지만 느낌이 다르다. 현재 Arphenotype은 디지털을 사용한 작업을 더 많이 하지만 아날로그 방식 또한 매우 중요하다고 생각한다. 스케치북이 내가 즉흥적으로 언제든지 사용할 수 있어 좋다. 몇 년이 지난 후 다시 볼 때에 그 안에 들어 있는 것들을 보며 자주 놀라곤 한다.

Any episodes or memories related to a notebook?

It may sound strange, but **I have the feeling that the best ideas are born, when I do not have access to the internet** - maybe we are all slaves of this digital cloud already. At the moment I am commuting every week between Cologne and Berlin - which is a 4 1/2 hours train ride, one way. Here in Germany we do not have internet yet on the trains. So it is like being free, time to focus on reading books and working. Here is my notebook central element of my mobile office setup.

수첩에 관련된 에피소드가 있다면 들려달라.

이상하게 들릴지 모르겠지만, **최고의 아이디어가 떠오를 때는 내가 인터넷을 사용하지 않을 때라는 느낌이 든다.** 어쩌면 우리는 모두 이미 디지털 클라우드의 노예일지도 모른다. 요즘 나는 편도 4시간 반 걸리는 기차를 타고 매주 퀼른에서 베를린까지 출퇴근 한다. 독일에는 아직 기차 안에 인터넷이 설치되어 있지 않다. 그래서 책도 읽고 일도 하면서 좀 더 자유 시간을 가질 수 있는 것 같다. 이럴 때 수첩은 내 모바일 사무실의 중심 요소가 된다.

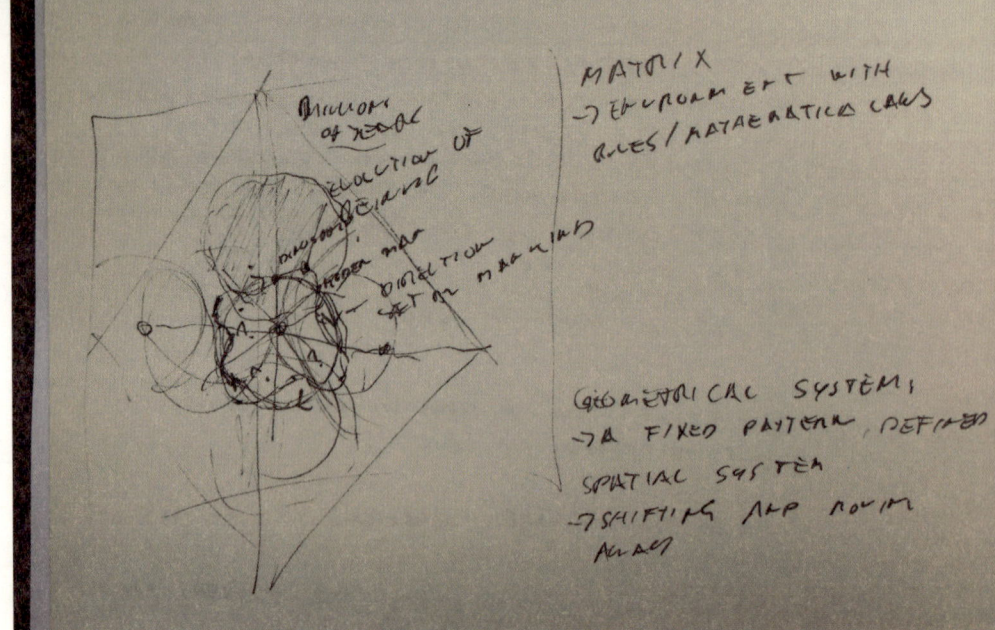

When and where do you use your sketchbook the most?

I use the notebook whenever it is suitable to hold ideas or input - specially at conferences, lectures or when I am invited as Guest Crit. At the beach, when I am reading a book to make references, to note quotes. When I am travelling, such as on planes, busses or in the train. Sometimes the best ideas appear in the weirdest situations, like waiting for the night bus after clubbing.

수첩을 가장 많이 사용하는 공간과 때는?

아이디어나 작업 한 것을 적어야 할 때 스케치북을 사용한다. 특히 컨퍼런스나 강의, 또는 게스트 크리틱으로 초대 받았을 때 많이 사용하는 편이다. 바닷가에서 책을 읽으면서도 적어둘 것이 있으면 사용하기도 한다. 비행기나 버스, 또는 기차를 타고 여행을 할 때에도 사용한다. 가끔 보면 가장 좋은 아이디어는 가장 이상한 순간에 떠오르기도 한다. 클럽에서 나온 후 나이트 버스를 기다리고 있을 때처럼 말이다.

I have the feeling that the best ideas are born, when I do not have access to the internet.

What influence does a sketchbook have in your projects and life as an architect?

The sketch book is the 2nd step within formulating a system - first step is always the brain. Sometimes an idea is born and rethought over weeks, before I start sketching. Then it even can rest for years in my sketchbook before I work further on it. Once it is imbedded in a 3D modelling program the sketch process is evaluated on tracing paper, working models etc. - not really in the sketchbook itself anymore - unless it is a theoretical project.

수첩은 당신의 작품과 삶에 어떤 영향을 끼치는가?

시스템의 첫 번째 단계는 뇌고, 두 번째 단계가 바로 스케치북이다. 아이디어가 처음 생각났을 때 바로 그리는 것이 아니라 몇 주 동안 생각하고 스케치를 시작한다. 그러다 또 더 작업을 하기까지 몇 년 동안 스케치북 안에 남아있기도 한다. 이론적인 프로젝트가 아닌 이상, 스케치가 3D 모델 프로그램 안에서 그려진 후에는 스케치북보다는 트레이싱지나 모형을 통해 아이디어를 발전시킨다.

Are there anything else other than a sketchbook that you use to keep a record of your thoughts and ideas?

Yes, they are all over the place. I have to admit, I am not an organized person with my ideas, sometimes I use books for sketches, napkins, physical models and so on - sometimes the original ideas is taken up at another location, being aware that the systematic idea stays the same.

수첩 외에 자신의 생각을 기록하는 방법과 도구는 무엇이 있는가?

여기저기 있다. 나는 아이디어들을 잘 정리해두는 편이 아니다. 가끔은 책이나 냅킨 또는 모형에 스케치를 할 때가 있다. 또 때로는 초기 아이디어의 체계적인 요소는 지키면서 그 아이디어 자체는 다른 곳에서 발전시키기도 한다.

PATTERNS ARE MESSAGES, HENCE WE CALL IT CARCACE.
LANGUAGE BECOMES TOUCHSTONE OF PERSONAL IDENTITY

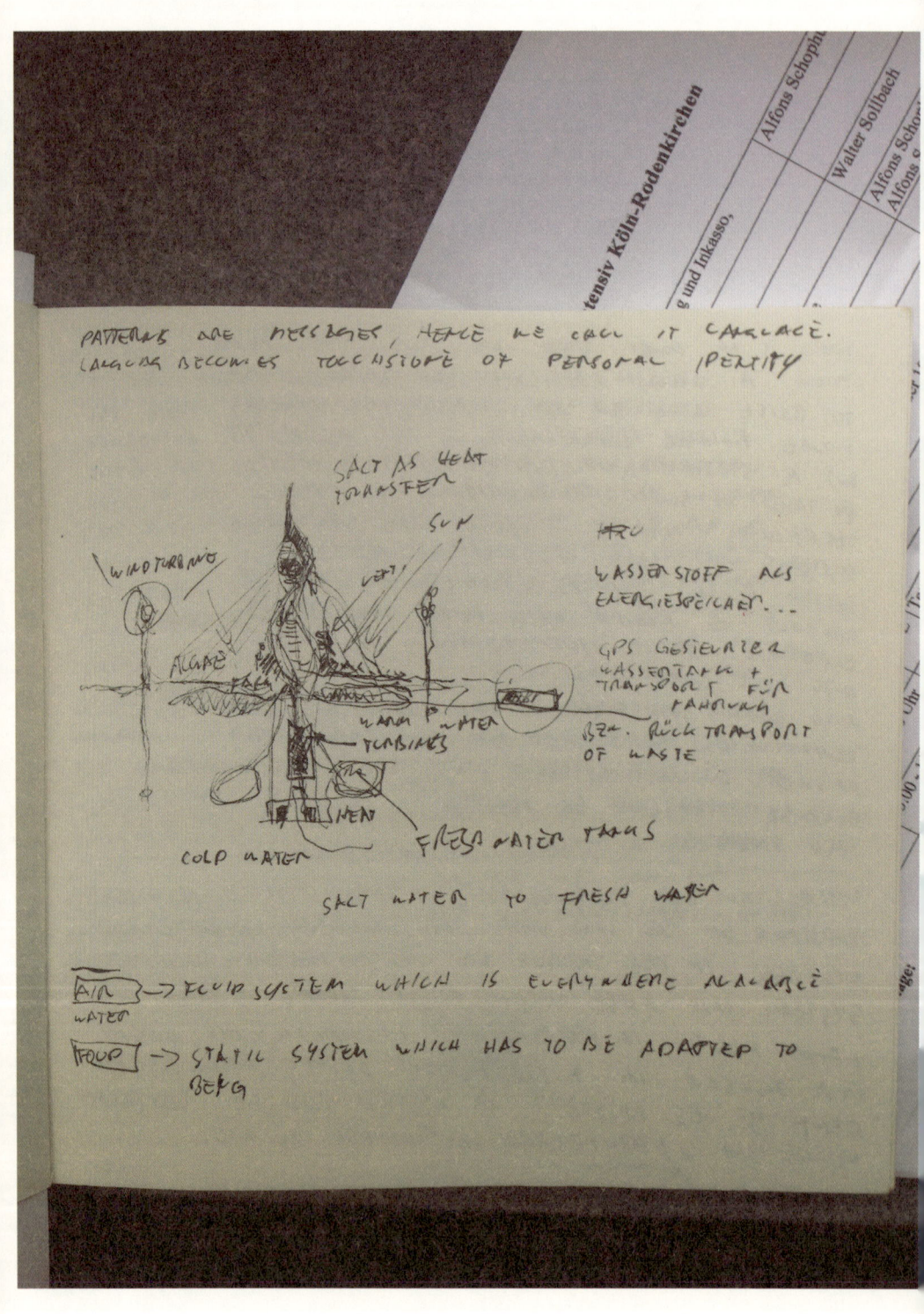

SALT WATER TO FRESH WATER

AIR → FLUID SYSTEM WHICH IS EVERYWHERE AVAILABLE
WATER

FOOD → STATIC SYSTEM WHICH HAS TO BE ADAPTED TO
          BERG

*Flo[at]wer
Design: ARPHENOTYPE

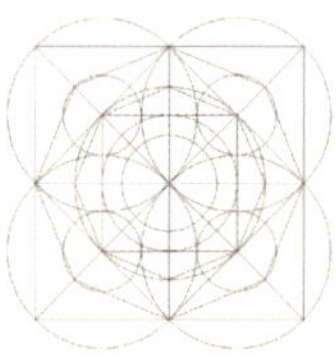

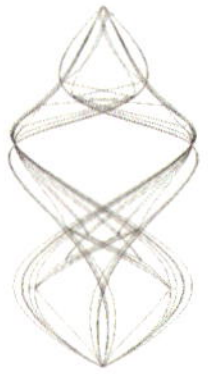

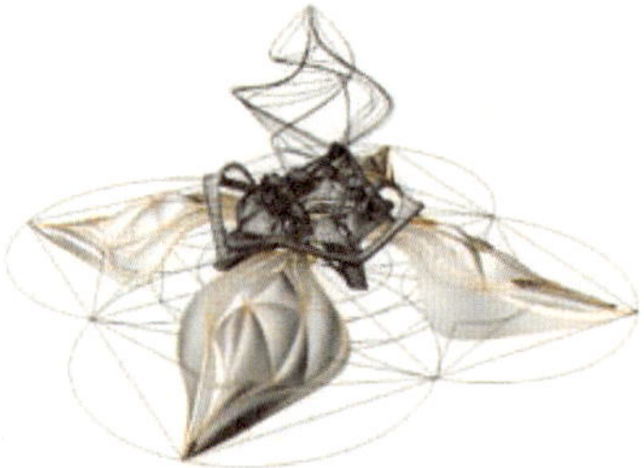

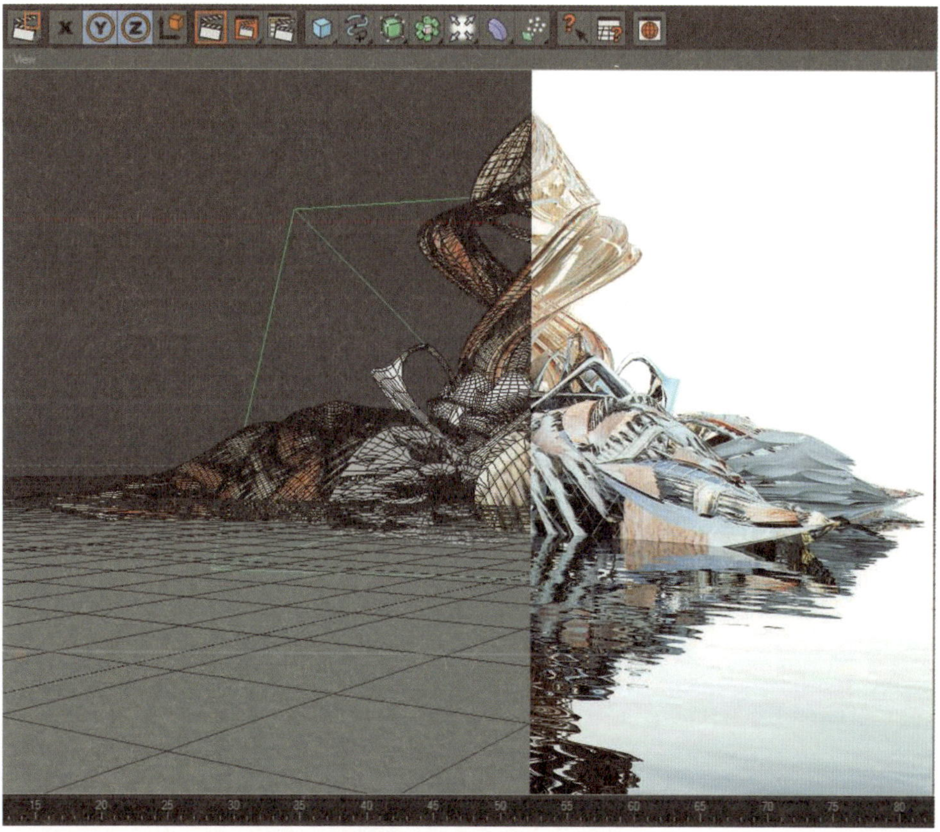

201

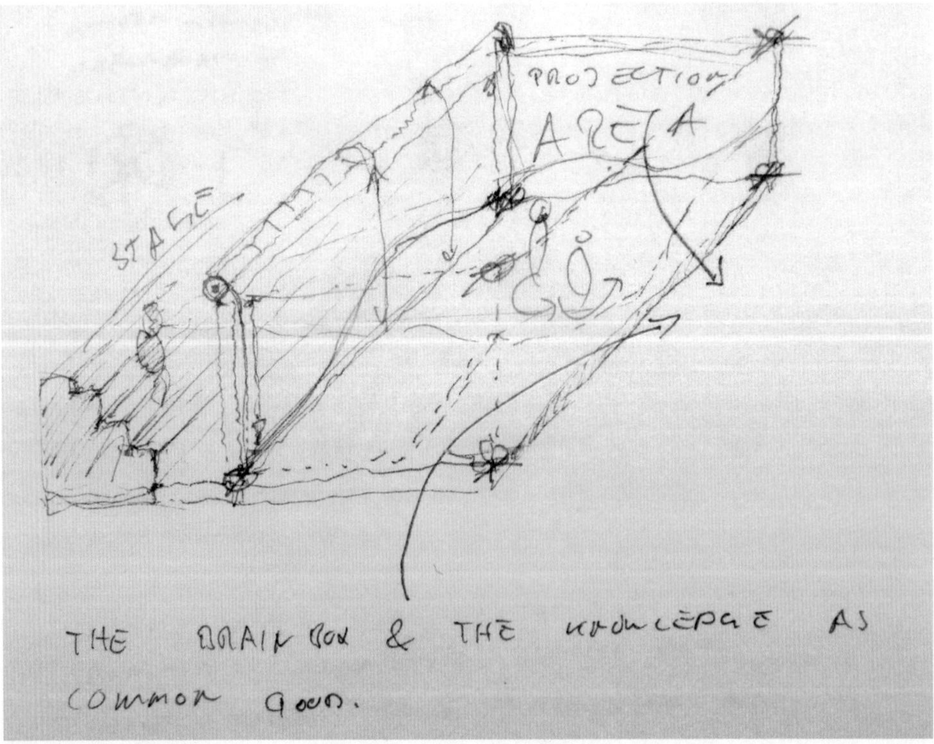

*FLoating Permaculture

THE BRAIN BOX & THE KNOWLEDGE AS COMMON GOOD.

projector / sensor 1
projector / sensor 2
projector / sensor 3
environment
environment
projector for environment

LIFT TO EMPTY SPACE
SYS I
SENSORS
SYS II
PERFORMANCE
SPACE
SYS III
SPACE
WIND
SYS IV
NATURAL SOURCES
• SUN
• WIND
• FOG (RAIN)
SUN
FOG

14.50
25.00

14.50

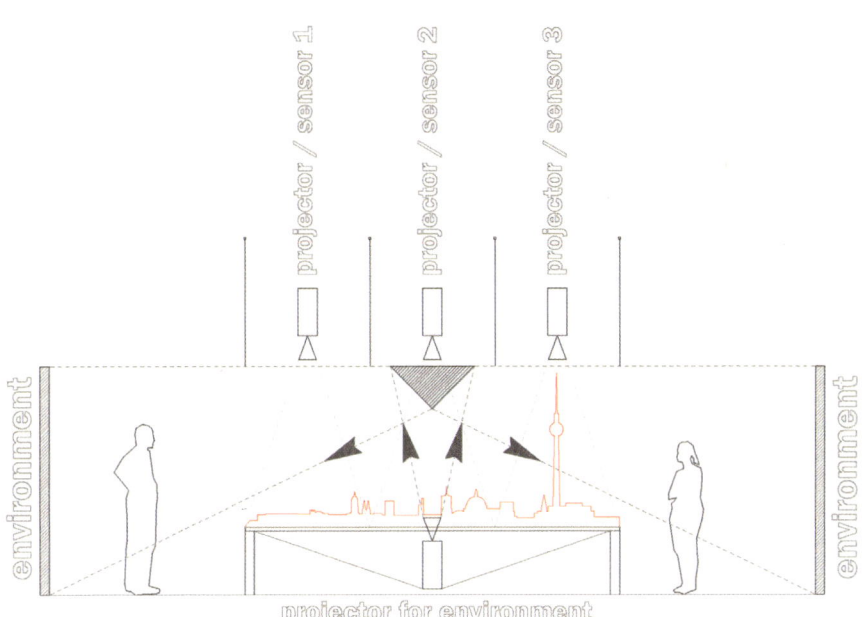

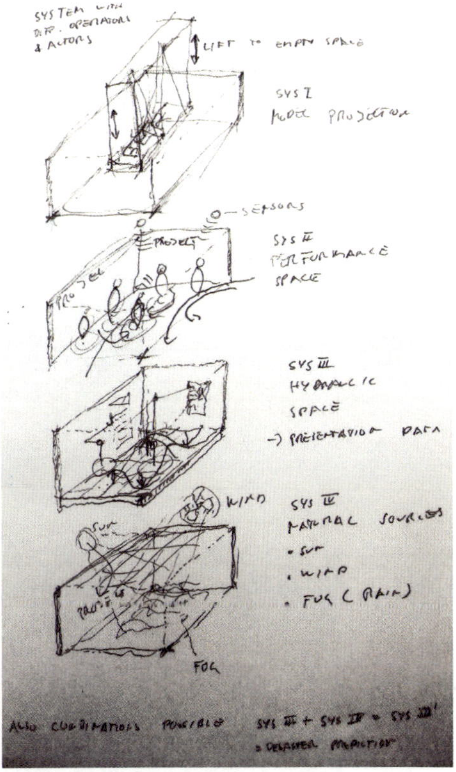

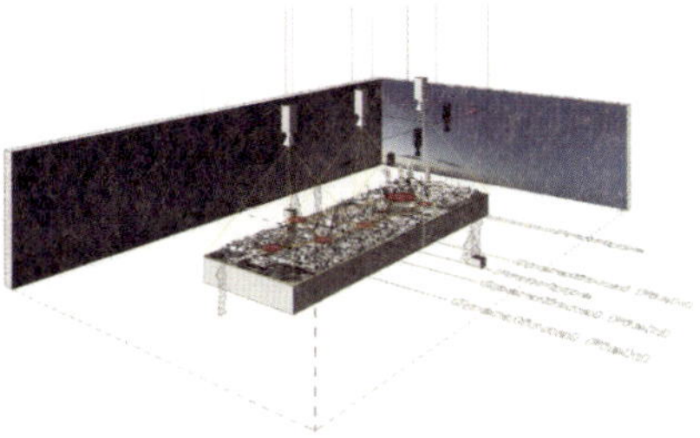

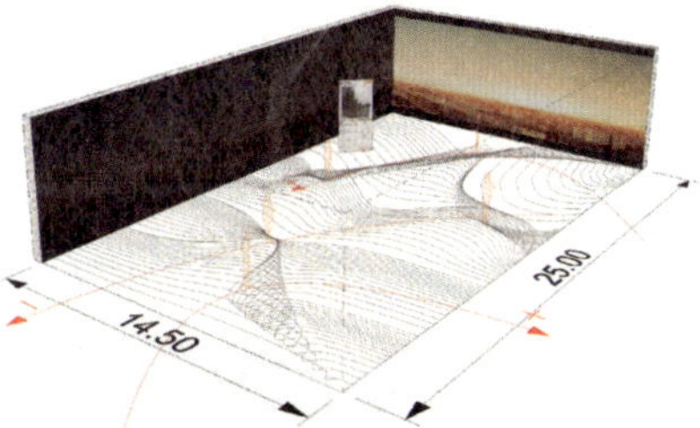

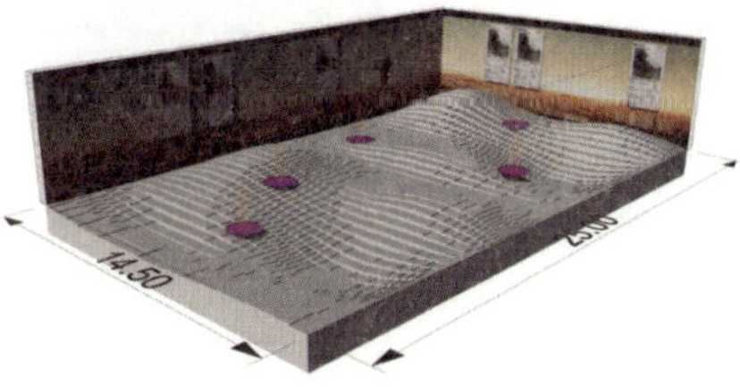

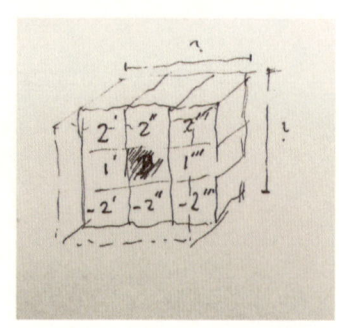
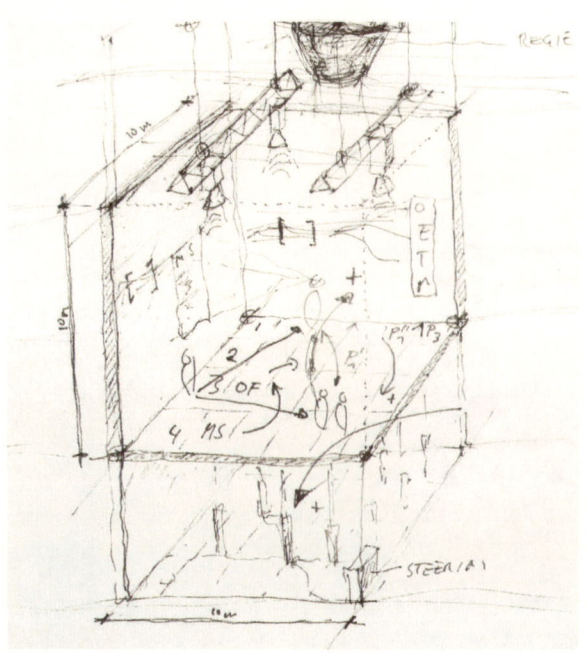
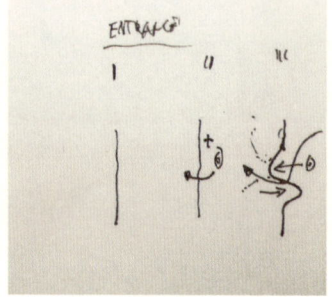
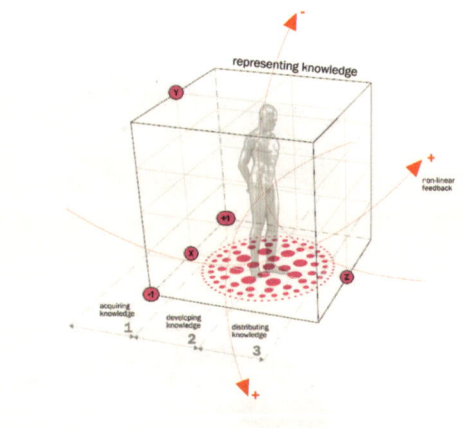
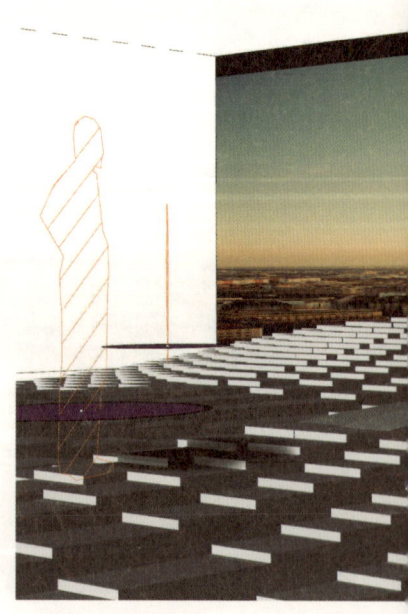
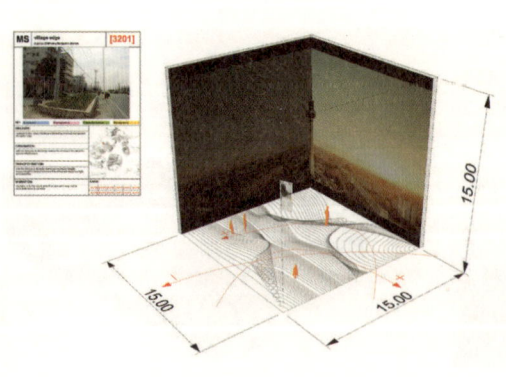

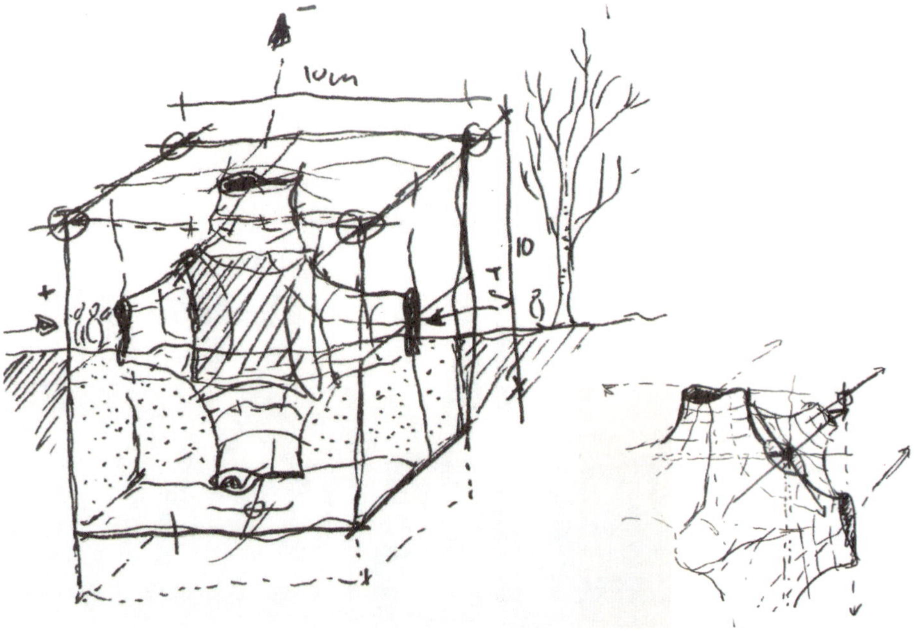

205

OSA

www.o-s-a.com

In 2002, Open Source Architecture (www.o-s-a.com) was founded by Chandler Ahrens (Washington University in Saint Louis), Eran Neuman (Tel Aviv University), and Aaron Sprecher (McGill University). As such, Open Source Architecture has been organized as an architectural group that undertakes projects ranging from industrial design, exhibition design to large urban installations and buildings. At Open Source Architecture, researchers and staffs are engaged in diverse expertise considering that architectural design is foremost linked to our current technological conditions. The main emphasis is placed on investigating new modes of spatiality and materiality made available through the accelerated changes occurring in our contemporary culture, technological and environmental conditions.

From experimental structures (The Hylomorphic Project, West Hollywood) in the field of structural optimization to large urban installations (ParaSolar, Tel Aviv), exhibition design (Evolutive Means, New York City) and residential buildings (Slrsrf, Culver City), Open Source Architecture focuses on the combinatory relation between the architectural object and its environment, being natural, artificial or even virtual. This approach to architecture recognizes the necessity to reconsider the nature of the architectural object as a dynamic system that stems from transdisciplinary collaborations with experts in structural and civil engineering, computer sciences, environmental design, and manufacturing among others. As a result, Open Source Architecture acts as a decentralized design and research organization with offices in Los Angeles, Tel Aviv and Montreal; and experts in Europe, Asia and the United States.

What is a notebook to you?

Our work at OSA is digitally based, developing complex geometry or computational systems of variable relationships. Therefore, the use of a sketching and the notebook has changed simultaneously as the design process has changed from 20 years ago when we were starting out in the design field. This is not to say that the notebook is any less important, but **rather that the information it provides to initiate or develop a project has changed.** Today, I find it more difficult to sketch out an entire building idea in plan, section or perspective since the quality of space due to the geometric complexity is difficult to establish in a drawing. If the design is less complicated, such as a chair with less geometry, I can still use the sketch to attempt to work out the main design ideas. But, typically I use sketches to diagram relationships between spaces or systems. There are two basic types of sketches I use as a design tool; analytical and relational. Analytical sketches either interpret existing conditions or project more specific qualities about a space or a series of objects while relational sketches attempt to map out variables in a system to speculate on the range of possible associations between components.

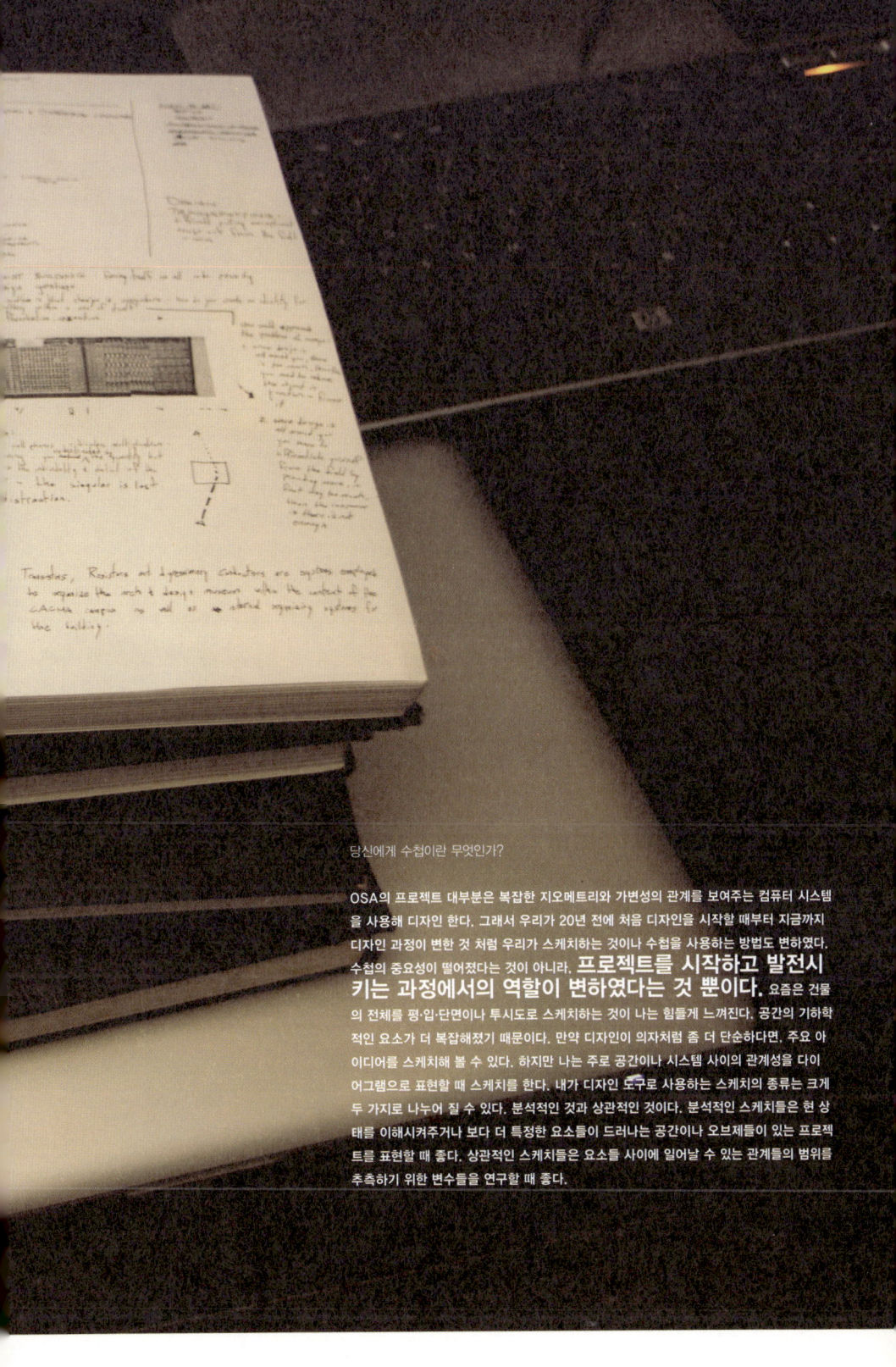

당신에게 수첩이란 무엇인가?

OSA의 프로젝트 대부분은 복잡한 지오메트리와 가변성의 관계를 보여주는 컴퓨터 시스템을 사용해 디자인 한다. 그래서 우리가 20년 전에 처음 디자인을 시작할 때부터 지금까지 디자인 과정이 변한 것 처럼 우리가 스케치하는 것이나 수첩을 사용하는 방법도 변하였다. 수첩의 중요성이 떨어졌다는 것이 아니라, **프로젝트를 시작하고 발전시키는 과정에서의 역할이 변하였다는 것 뿐이다.** 요즘은 건물의 전체를 평·입·단면이나 투시도로 스케치하는 것이 나는 힘들게 느껴진다. 공간의 기하학적인 요소가 더 복잡해졌기 때문이다. 만약 디자인이 의자처럼 좀 더 단순하다면, 주요 아이디어를 스케치해 볼 수 있다. 하지만 나는 주로 공간이나 시스템 사이의 관계성을 다이어그램으로 표현할 때 스케치를 한다. 내가 디자인 도구로 사용하는 스케치의 종류는 크게 두 가지로 나누어 질 수 있다. 분석적인 것과 상관적인 것이다. 분석적인 스케치들은 현 상태를 이해시켜주거나 보다 더 특정한 요소들이 드러나는 공간이나 오브제들이 있는 프로젝트를 표현할 때 좋다. 상관적인 스케치들은 요소들 사이에 일어날 수 있는 관계들의 범위를 추측하기 위한 변수들을 연구할 때 좋다.

Any episodes or memories related to a notebook?

I was moving my family from the east coast of the United States to the west coast (Los Angeles) and we drove across the country, stopping in Utah to see some amazing rock formations. I created a series of analytical sketches to interpret and project a sense of scale to the fairly ambiguously scaled, but massive objects. In addition, I had brought a child's toy camera similar to a Polaroid camera that could create instant photos, but they were only 3 x 4cm. I pasted the photos of the rock formations in the sketchbook to establish a dialog between the photographic recording and my analytical interpretation. I was not interested in the accuracy of how the rocks existed since the photo served that purpose. **I was more interested in the qualities of scale, mass, and hierarchy that I could project onto them through the process of sketching.**

When and where do you use your notebook the most?

I mostly sketch at my desk, though it is not always in my sketchbook. Therefore, I use my sketchbook half of the time at my desk and the other half when I travel since its portability makes it easy to carry.

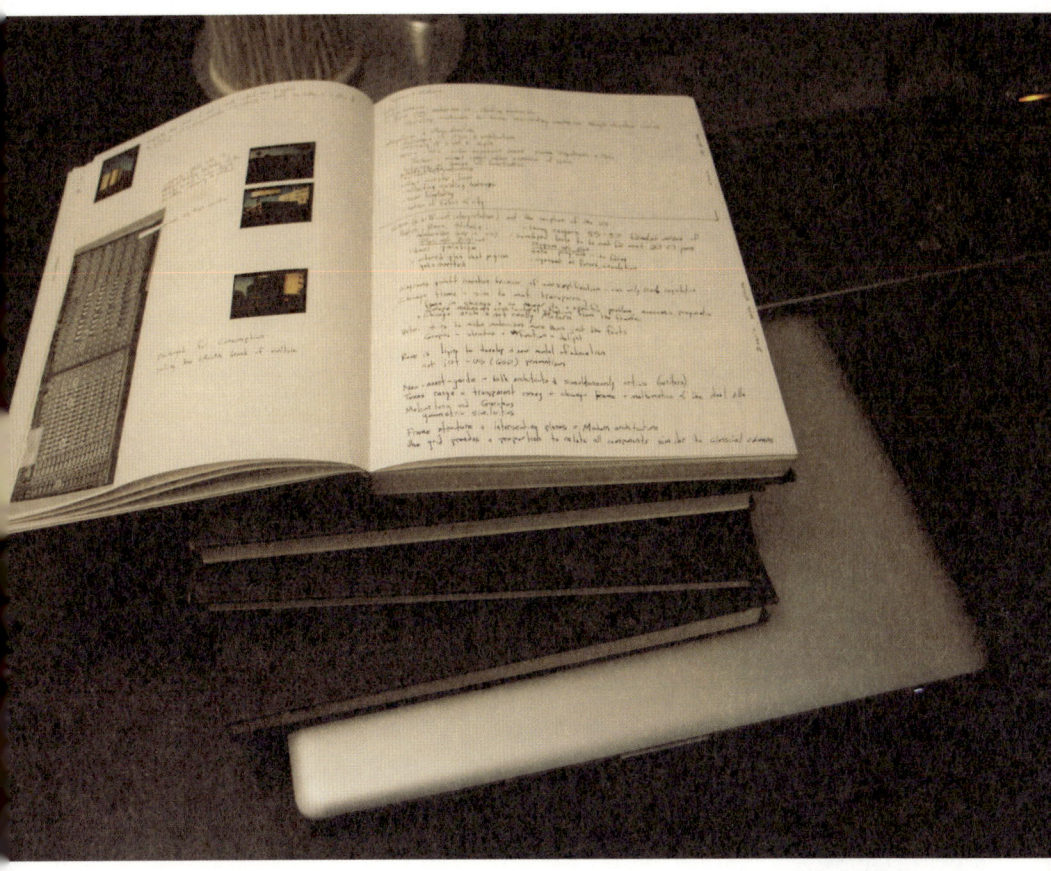

수첩에 관련된 에피소드가 있다면 들려달라.

전에 우리 가족 모두 미국 동부에서부터 서부(로스앤젤레스)까지 운전해서 이사를 하면서 놀라운 암석 형성들을 보기 위해 유타에 들른 적이 있다. 그때 거대하지만 꽤 애매모호하기도 한 규모의 암석들을 이해하고, 그 크기들을 가늠하기 위해 분석 스케치를 했었다. 그리고 폴라로이드 카메라와 비슷했지만 사진 크기가 3×4cm 밖에 안되는 아이의 토이 카메라를 가지고 있었다. 암석 사진을 찍은 후 스케치북에 붙여 나의 분석적인 해석과 사진 기록간의 관계를 형성시켰다. 사진이 있었기 때문에 나는 얼마나 정확하게 암석을 그렸는지는 상관하지 않았다. **나는 스케치를 통해 표현할 수 있는 스케일과 매스 그리고 위계질서에 더 관심 있었다.**

수첩을 가장 많이 사용하는 공간과 때는?

나는 주로 책상에서 스케치를 하지만 항상 스케치북에 하는 것은 아니다. 그래서 내가 스케치북을 사용하는 시간의 반은 책상에서, 그리고 다른 반은 휴대하기 간편해서 주로 여행을 다니면서 한다.

I was more interested in the qualities of scale, mass, and hierarchy that I could project onto them through the process of sketching.

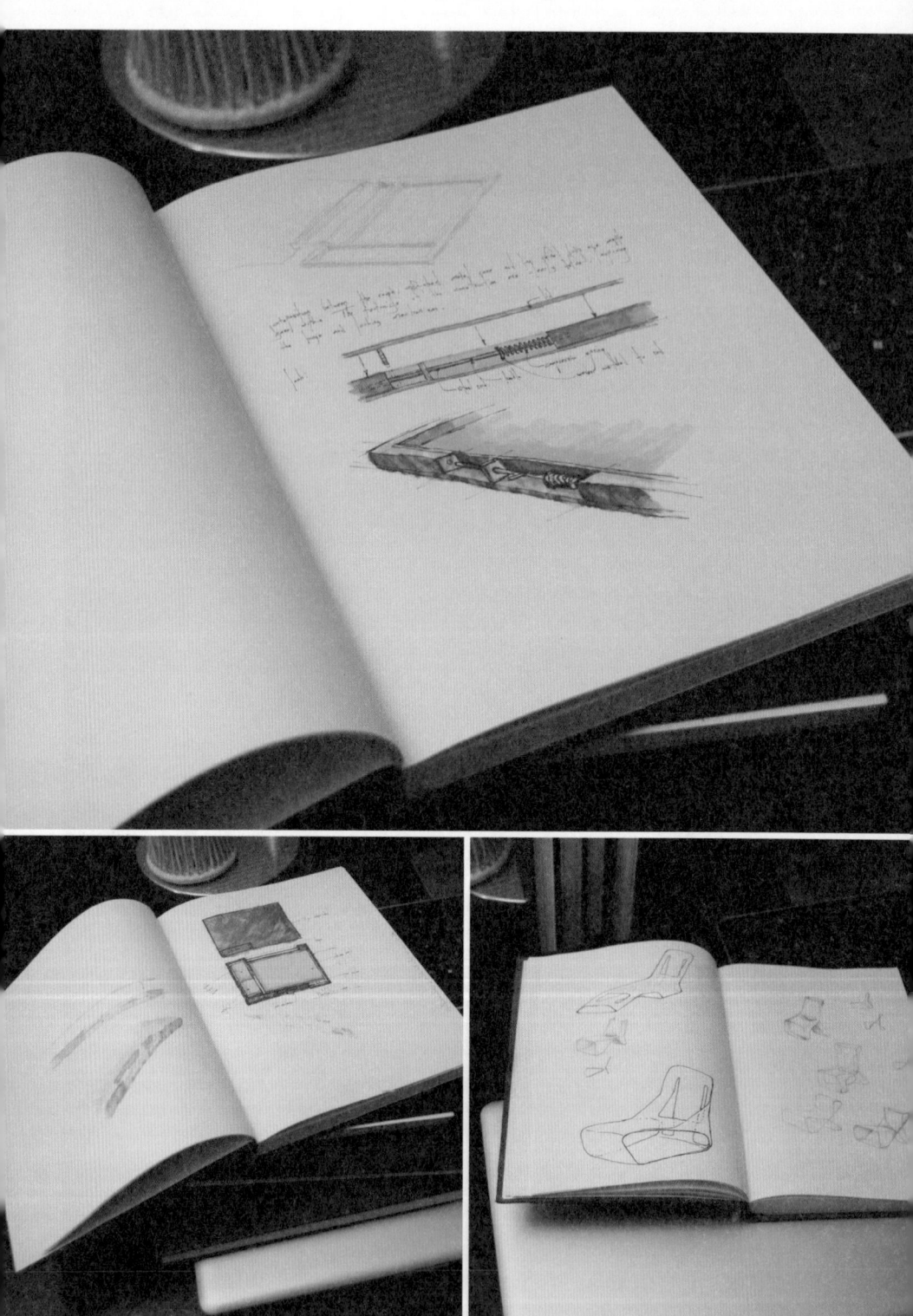

What influence does a notebook have in your projects and life as an architect?

Even though our design process focuses on digital methods, I find sketching to be an essential design tool from establishing initial ideas through the project development and into construction. Sketching is essential as a way for me to measure my own design decisions as well as a means to communicate ideas quickly and visually to others. My notebook in particular, is valuable for me to evaluate and measure initial design ideas that are too young to be fixed where the real value of the sketch relies on the inexact qualities that suggest a range of possible interpretations. It is within these potential interpretations that creative ideas cultivate.

수첩은 당신의 작품과 삶에 어떤 영향을 끼치는가?

비록 우리의 디자인 과정은 디지털 방법에 초점을 두고 있지만, 처음 아이디어를 내고 프로젝트를 발전시키기 위해서 스케치는 **필수적인 디자인 도구**라고 생각한다. 스케치는 나 자신의 디자인 결정들을 다시 보게끔 해줄 뿐만 아니라, 아이디어들을 다른 사람들과 시각적으로 빨리 소통할 수 있도록 해주는 것이다. 특히 나의 수첩은 확정짓기를 이른 초기 디자인 아이디어들을 평가하고 조정할 수 있도록 해주기 때문에 나에게는 매우 소중한 것이다. 그리고 스케치의 진정한 가치는 그 부정확한 속성으로부터 나온다고 생각한다. 넓은 범위 내에서 다양한 해석을 할 수 있도록 해주기 때문이다. 이러한 **잠재된 해석 속에 창의적인 아이디어들이 생기는 것이다.**

Are there anything else other than a notebook that you use to keep a record of your thoughts and ideas?

I sketch quite often, though it may not seem like I do because I don't always use my notebook. I sketch a lot on post-it notes and random pieces of loose paper. Really, I sketch on **whatever is closest to me** when I have a thought. The problem is since those sketches are not bound in a book, they tend to get lost after a while or when I move onto the next project. They should be kept and pasted into a notebook since it is a record of your thought process.

수첩 외에 자신의 생각을 기록하는 방법과 도구는 무엇이 있는가?

수첩을 항상 사용하지 않을 뿐이지, 나는 자주 스케치를 한다. 주로 포스트잇 종이나 일반 용지에 많이 하는 편이다. 주로 생각이 났을 때 나에게 **가장 가까운 아무것**에나 스케치를 한다. 문제는 한 책으로 제본되어 있는 것이 아니기 때문에 시간이 지나거나 다음 프로젝트로 넘어 갈 때 잃어버리기 십상이다. 생각 과정을 기록해 놓는 것이기 때문에 수첩에 붙여 보관하는 것이 좋다.

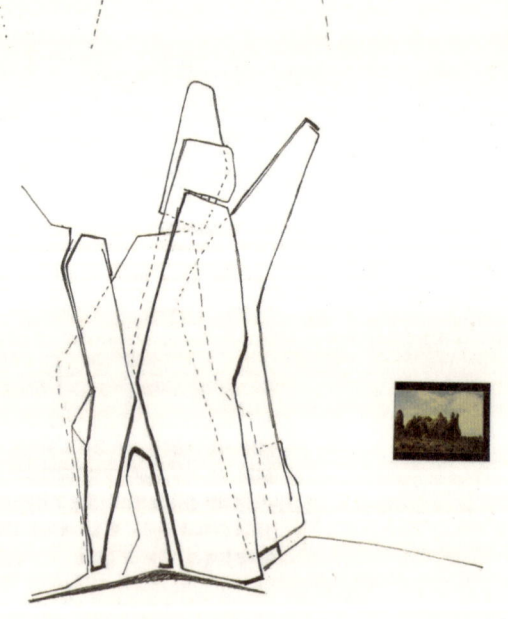

It is within these potential interpretations that creative ideas cultivate.

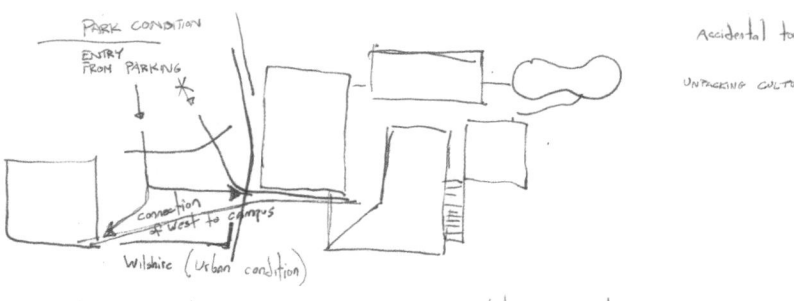

- activate the Wilshire facade by pushing circulation to it

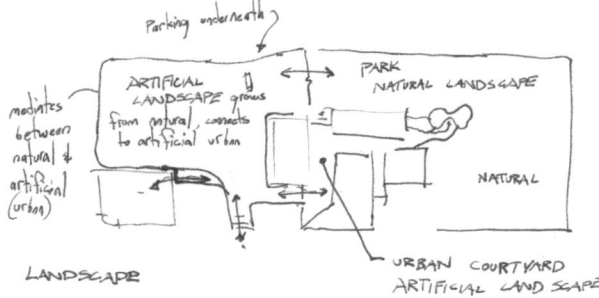

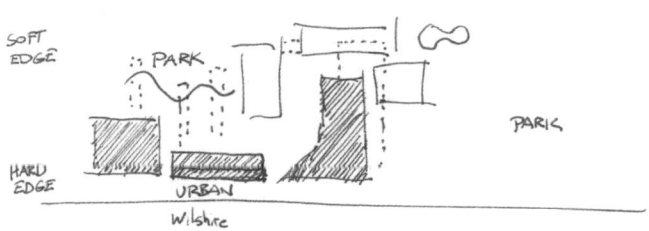

DISTRACTED ATTENTION

TRANSISTORS, RESISTORS & DIGRESSIONARY CONDUCTORS

TRANSISTORS — ICONIC IMAGE DEFINITION
1. TECTONIC MASS — DIAGRAM
2. GROUND PLANE — DIAGRAM — MIES NAT GALL — AIG SECTION
   PARKING & HARD
   PROGRAM UNDERNEATH

RESISTOR — ICONIC IMAGE DEFINITION
1. EXTERNAL — SIGN / IDENTITY DIAGRAM — MAY CO — MASTODON — FOOTBALL HALL OF FAME
   THRESHOLD
2. INTERNAL — NODES / POINTS OF INTEREST / DEFLECTORS — DIAGRAM

DIGRESSIONARY CONDUCTORS — ICONIC IMAGE DEFINITION
A. DISTRACTION PARADOX
1. DIAGRAM — FROM PARKING TO MAIN CAMPUS — C+RB - CARPENTER CENTER
   DIAGRAM — EVENING PARKING
2. EXTERIOR — DIAGRAM SHOWING EXHIBITS FROM AUTO — IMAGES FROM INSIDE CAR OF PASSING STOREFRONTS
   INTERIOR — DIAGRAM SHOWING — JUSSIEU - KOOLHAAS
   SITUATIONIST DRIFT
JUMPERS

differentiation through grotesque
sameness

PHOTOS TO GET:
MAY CO
MASTODON
~~FOOTBALL HALL OF FAME~~
~~STOREFRONTS FROM CAR~~
JUSSIEU - KOOLHAAS
etc

DESIGN TRAINSPOTTING —
difficult pulling exceptional design out from the field of norm

EGOIST BUILDING - forcing itself on all who pass by
Design
The problem is that design is everywhere - how do you create an identity for
something within a sea of itself?
differentiation; opposition

One could approach the problem 2 ways:
1. since design is all around you, there is too much, therefore you need to reduce the object in question — framing it
2. since design is all around you, you have to differentiate yourself from the field by standing more, in fact way too much, then the response is there is not enough

Differentiation:
10,000 cell phones; ridiculous multiplication —
overwhelming - you notice are distracted by the quantity but
as miss the individuality & detail of the
singular — the singular is lost in distraction.
Identity

Transistors, Resistors and digressionary conductors are systems employed to organize the arch & design museum within the context of the LACMA campus as well as an internal organizing systems for the building.

 iconography, sign, marker of what is inside without having to explain. the gesture is aimed at passing automobiles in an attempt to identify the building as a place of commerce

iconography - although kitch - it is effective at identifying the history of the tar pits by using a common notion of time effects presumed by the artifacts

recognizable only through association

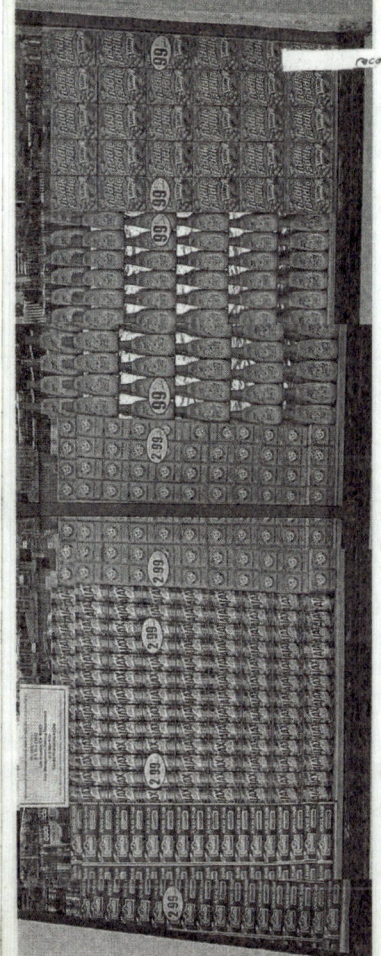

Packaged for Consumption
selling the LACMA brand of culture

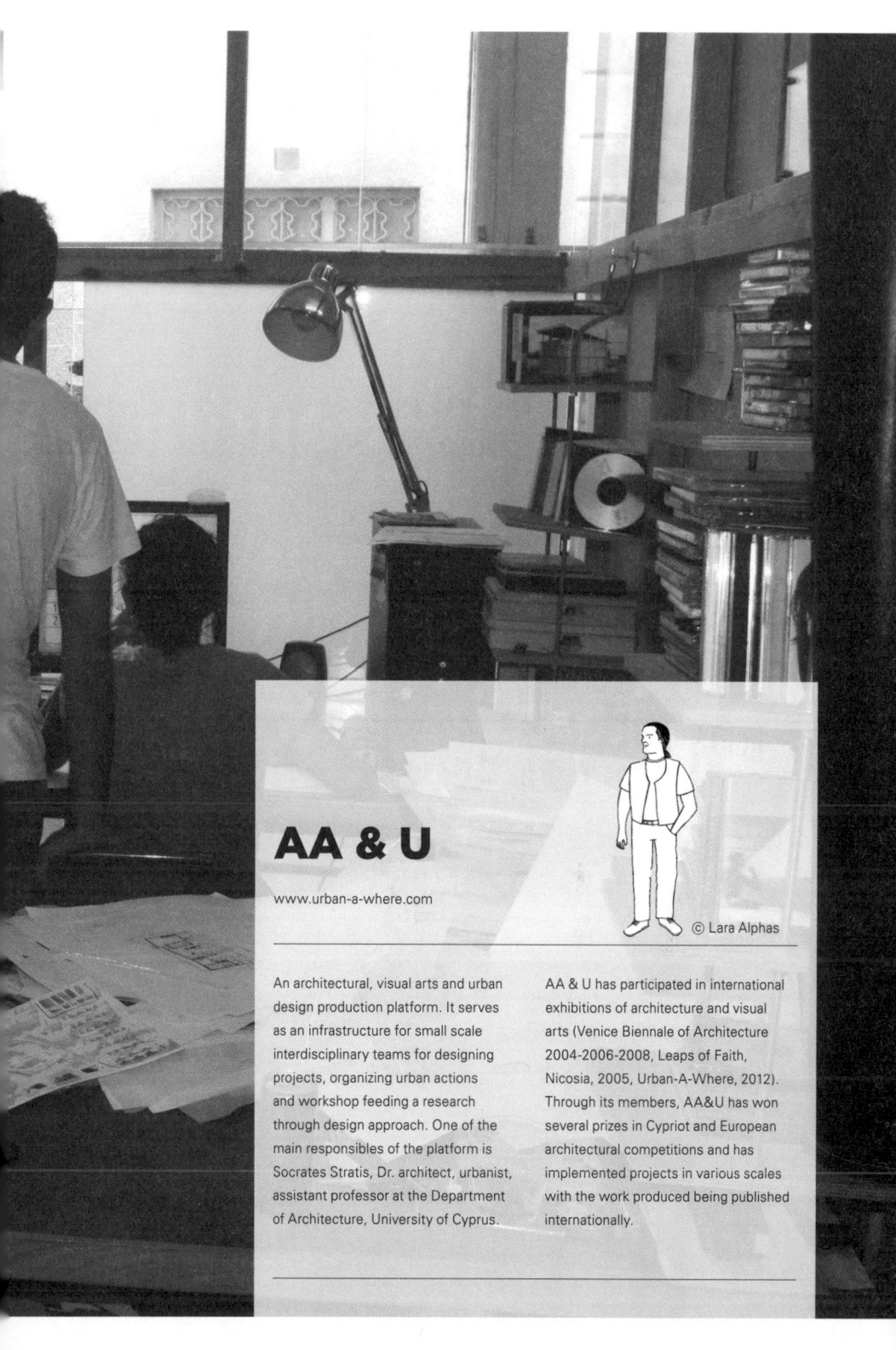

AA & U

www.urban-a-where.com

© Lara Alphas

An architectural, visual arts and urban design production platform. It serves as an infrastructure for small scale interdisciplinary teams for designing projects, organizing urban actions and workshop feeding a research through design approach. One of the main responsibles of the platform is Socrates Stratis, Dr. architect, urbanist, assistant professor at the Department of Architecture, University of Cyprus.

AA & U has participated in international exhibitions of architecture and visual arts (Venice Biennale of Architecture 2004-2006-2008, Leaps of Faith, Nicosia, 2005, Urban-A-Where, 2012). Through its members, AA&U has won several prizes in Cypriot and European architectural competitions and has implemented projects in various scales with the work produced being published internationally.

interviewee: Socrates Stratis

What is a notebook to you?

My notebook is my mind. By the time something is imprinted in the notebook has the possibility to be remembered and to go through a process of transcription into design. The process of imprinting an idea in the notebook becomes an integral part of the design process. Everything is noted in order to allow a return to it, a recollection or a reconnection to other current thoughts. Such thoughts may take the form of a drawing, or the form of a text. Such thoughts may have a documentary character, a sort of mapping of actual conditions which I find either in physical environment, in books or in presentations of architects' work. In other cases, they may have a projective character, meaning they become an imprint of possibilities projected in the future. That could relate to initial thoughts about projects, such as the sketches attached relating to an architectural competition for the re-conversion of a former agricultural farm into an incubator of activities.

인터뷰이: 소크라테스 스트라티스

당신에게 수첩이란 무엇인가?

나의 수첩은 **나의 생각**이다. 수첩에 그려지는 순간 디자인으로 변화하는 과정을 밟아갈 가능성이 높다. 수첩에 아이디어를 새기는 것은 디자인 과정에 있어 필수적인 요인이다. 다시 되돌아가 생각도 해보고, 다른 아이디어들과도 비교하며 생각해보고 연결해볼 수 있도록 모든 것을 기록한다. 이러한 아이디어들은 그림이 될 수도 글이 될 수도 있다. 물리적인 환경이나 책, 또는 건축가들의 프레젠테이션을 보면서 마치 다큐멘터리를 기록하듯이 순간 순간의 상황을 매핑하기도 한다. 또 다른 경우에는 투사적이어서 미래에 실현 가능한 아이디어들의 시작점이 되기도 한다. 프로젝트 초기에 했던 생각들과 연관된 스케치들이다. 하나의 예로는 과거의 농장을 다양한 활동이 일어나는 곳으로 변화시키는 건축 공모전에 참가 했을 때 그렸던 스케치들이 있다.

Any episodes or memories related to a notebook?

I have realized that my notebook is my mind, when I went to a lecture by a fellow architect and I had forgotten to take it with me. During the whole presentation I was imagining how I was imprinting ideas in my notebook, starting off a process of transcription. **My notebook has become a valuable interface between myself** and the world allowing the projection of imaginaries and their link to any kind of precedence.

수첩에 관련된 에피소드가 있다면 들려달라.

동료 건축가의 강의에 가면서 나의 수첩을 잊어버리고 놓고 간 적이 있다. 나는 그때 내 수첩이 곧 나의 생각이라는 것을 깨달았다. 나는 수첩에 아이디어들을 어떻게 그려내고 있을까를 강의 내내 상상했었다. **나의 수첩은 나와 세상을 연결 시켜주는 중요한 접점이 되었다.** 상상했던 것들과 선례와 연결된 고리들을 가시화 시켜준다.

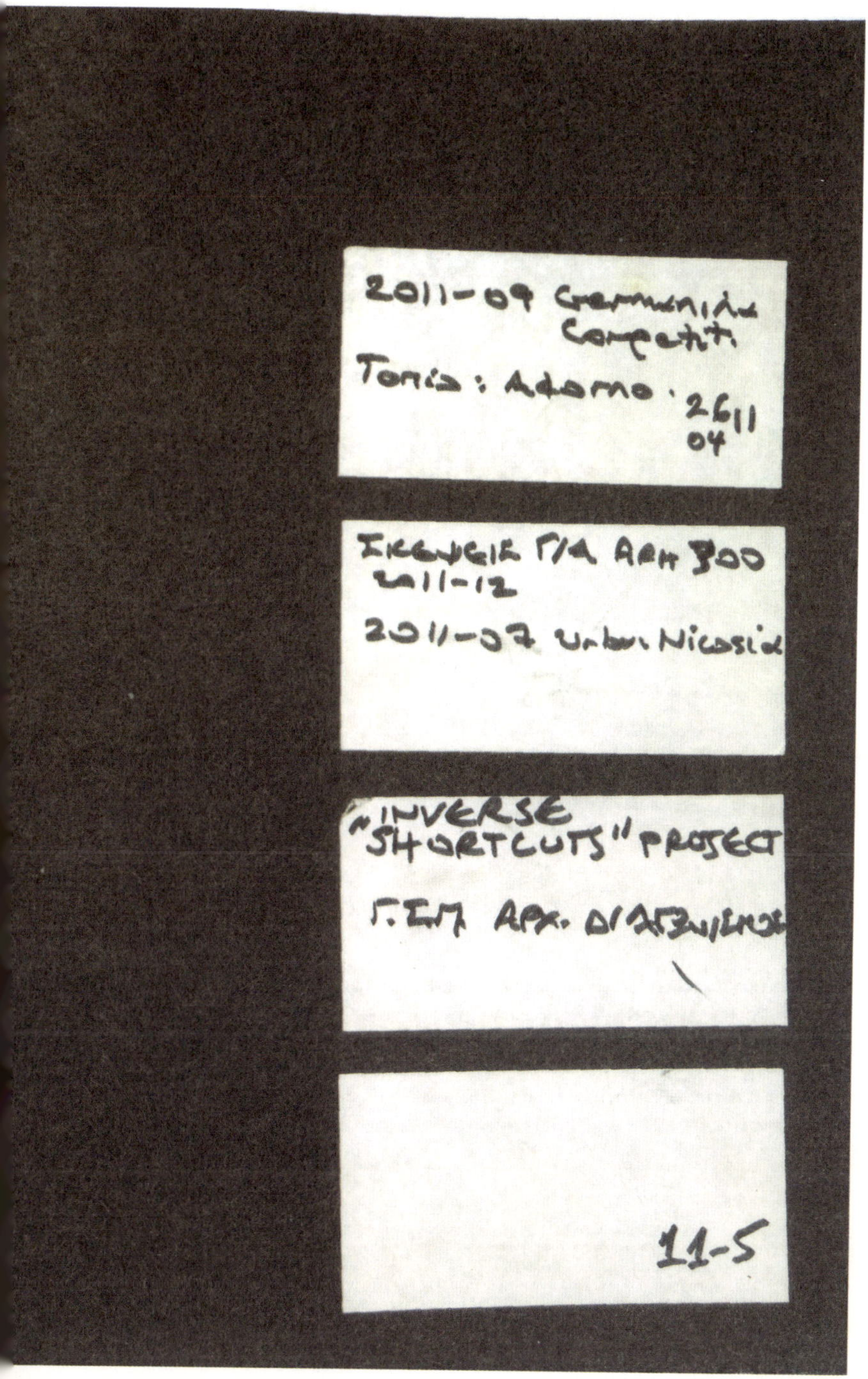

2011-09 Germanium
            Competiti
Tonia: Adamo · 26|11
                    04

ΙΚΕΝGIR Γ/α ΑΡΗ 300
2011-12
2011-07 Urban Nicosia

"INVERSE
 SHORTCUTS" PROJECT

Γ.ΕΜ ΑΡΧ. ΟΙ ΔΣΖΝ/ΕΝΩΕ

11-5

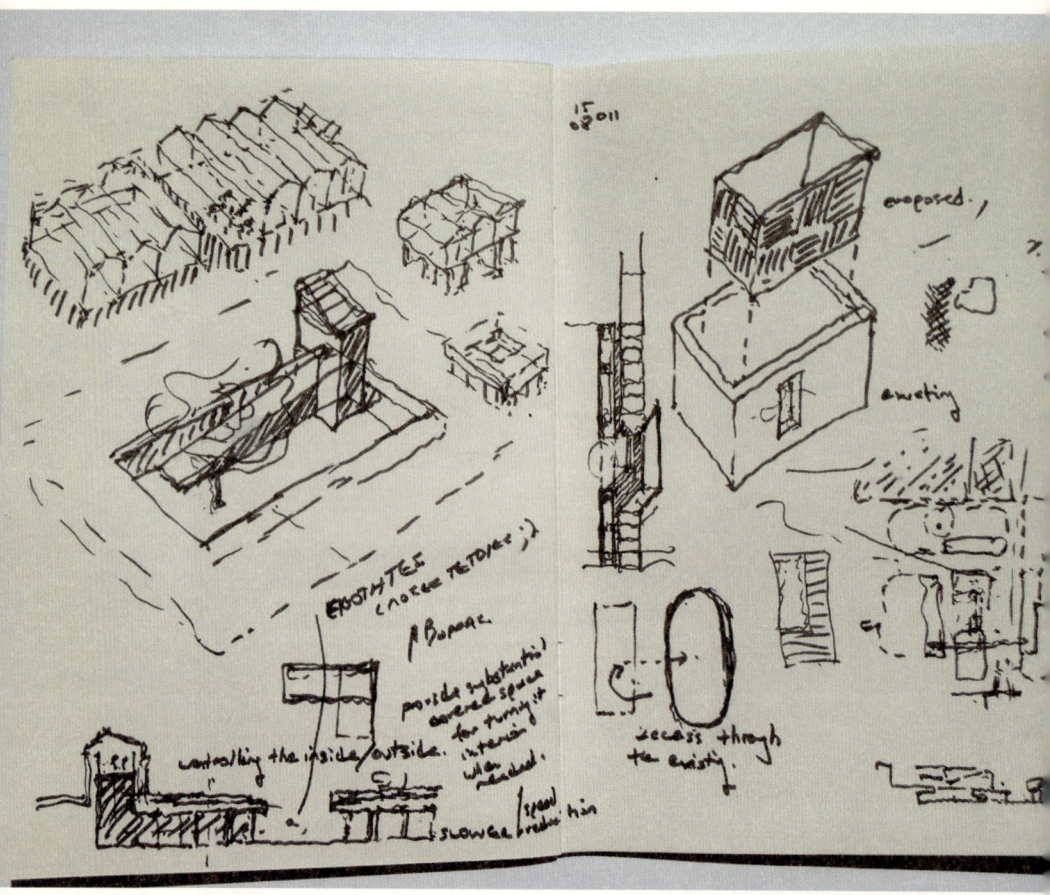

When and where do you use your notebook the most?

My notebooks have become over the last years very light so I can carry them in my backpack wherever I go. Usually I have more than one at any given time and they are organized around anything that I do, with the front cover being a sort of an index of such activities. During the last years, both drawing and writing, have become equally important in filling the pages. Quite often they are scanned in the digital environment becoming **the base** for design processes or writings about architecture and urban design.

My notebook has become a valuable interface between myself.

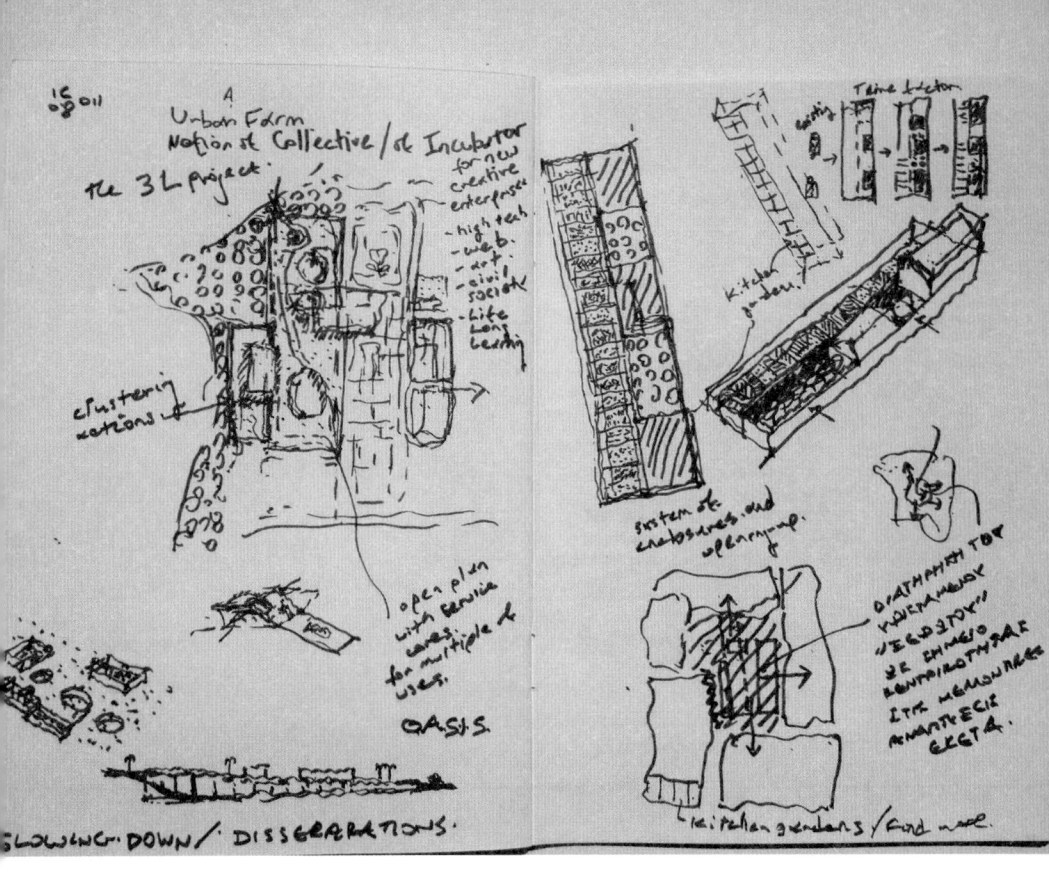

수첩을 가장 많이 사용하는 공간과 때는?

어디를 가든지 가방에 넣어 다니기 위해 지난 몇 년 동안 나의 수첩들은 점점 더 가벼워졌다. 나는 언제나 하나 이상 가지고 다닌다. 내가 하는 모든 활동마다 각각의 수첩이 있는데, 이를통해 정리 한다. 지난 몇 년동안 그리는 것과 쓰는 것 둘 다 페이지를 채우는데 사용되었다. 이 페이지들은 주로 스캔을 통하여 디지털 환경에서 디자인 하는 것과 건축이나 도시 디자인에 대하여 글을 쓰기 시작할 때의 **밑바탕**이 된다.

During the last ten years my involvement with architecture has been both through drawing and writing.

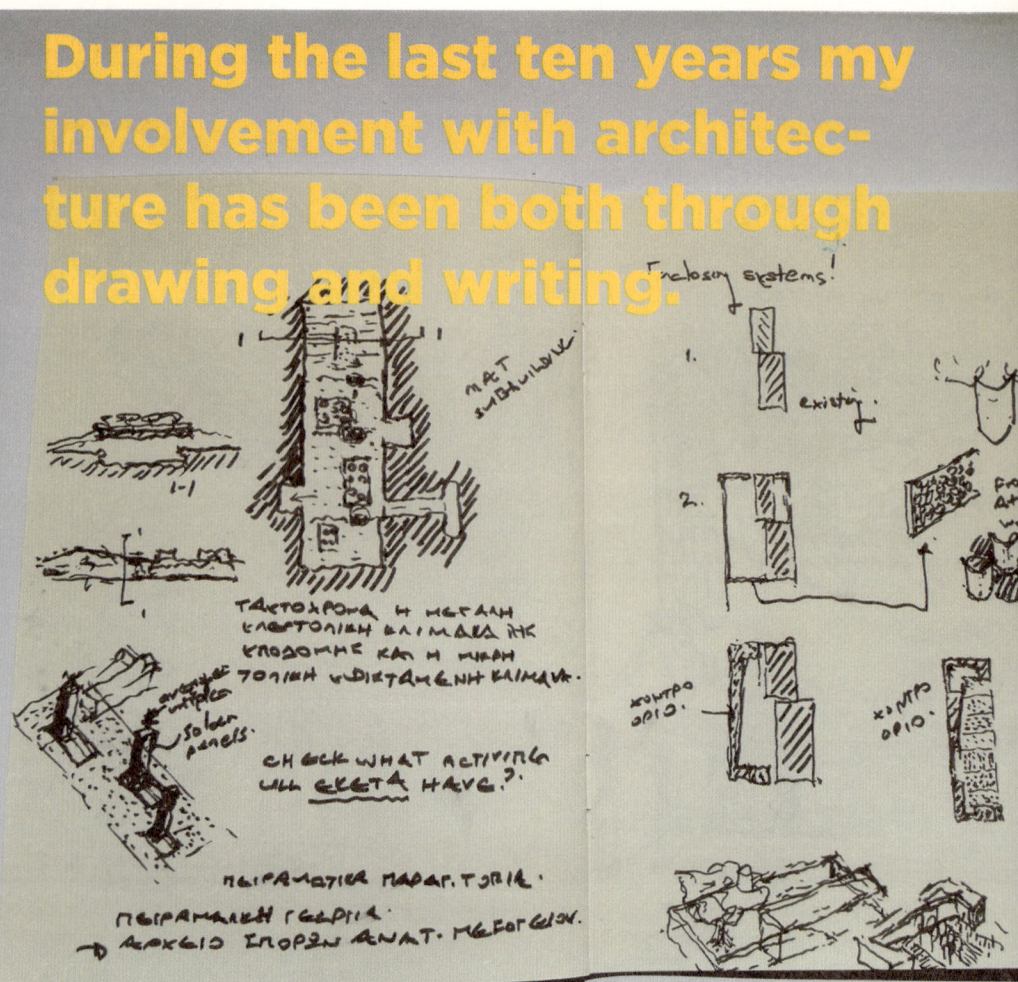

What influence does a notebook have in your projects and life as an architect?

During the last ten years my involvement with architecture has been both through drawing and writing. The sketchbook as I mentioned already, becomes a sort of register of thoughts during the different stages of projects which could have a drawing or a writing form. I always carry a backpack with me so I could put inside my notebook.

수첩은 당신의 작품과 삶에 어떤 영향을 끼치는가?

지난 10년 동안 나는 그리는 것과 쓰는 것을 통해 건축을 해왔다. 내가 앞서 얘기한 것처럼, 나의 스케치북은 프로젝트 진행 과정에서 나의 생각들을 정리해주는 곳이다. 그 생각들은 그림일 수도, 글일 수도 있다. 스케치북을 넣어 다닐 수 있도록 항상 가방을 들고 다닌다.

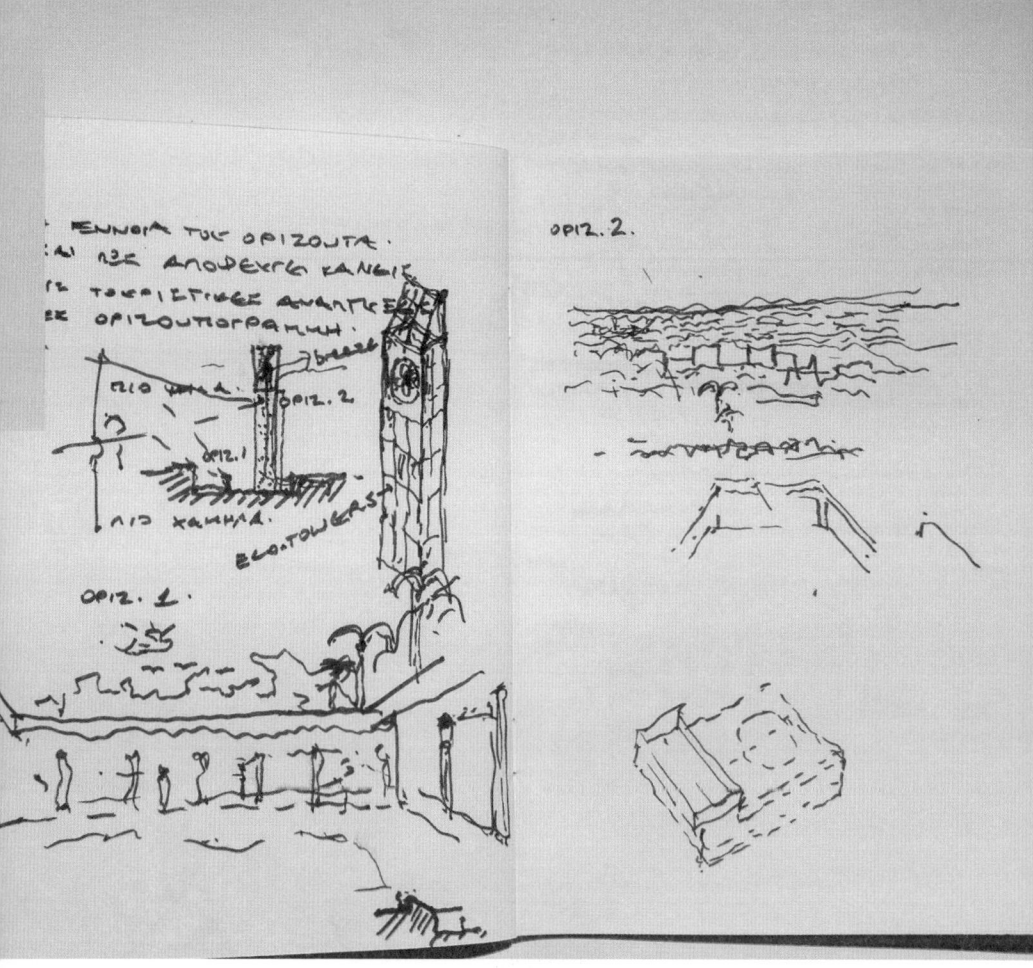

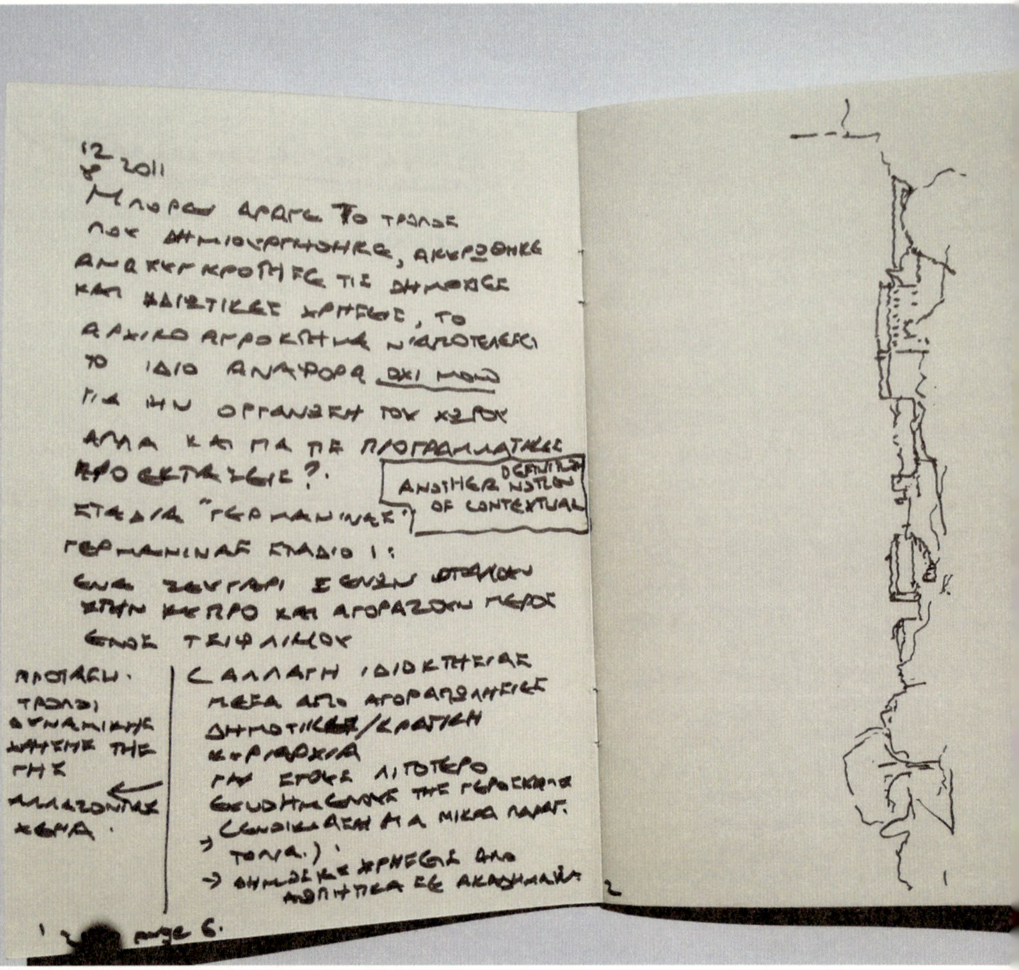

*European Competition for the Redevelopment of a former farm "Germanina", Cyprus
Collaborator: Anastasia Angelidou, architect
Assistants: Stavri Yiannakou(architect), Savvas Anastasiou(architect student),
Vicky Theodorou(architect student), Filippo Gur(architect student)

Are there anything else other than a notebook that you use to keep a record of your thoughts and ideas?

There are two other ways of keeping records of thoughts and ideas. The first is a specific about design process and the second one is generic regarding references of all sorts. The former is about using physical models as records of ideas about specific projects. Series of models from the very beginning of the design process materialize design ideas. Quite often it is one model that thanks to its easy manipulation is materialized out of a continuous overlaying of gestures during the various stages of the design processes. The second way of keeping records is through digital archives of images from travels in various cities. They are well registered so I could have easy access to them when I feel there is something to find during a design process.

수첩 외에 자신의 생각을 기록하는 방법과 도구는 무엇이 있는가?

생각과 아이디어들을 기록 할 수 있는 다른 두 가지의 방법이 있다. 첫 번째는 디자인 과정에 사용되는 것이고, 두 번째는 과정의 전반에 걸쳐 사용되는 것이다. 전자는 특정한 프로젝트 아이디어들을 모형을 만들어 기록을 남기는 것이다. 디자인 과정의 초기 단계부터 시리즈로 만들어진 모형들은 디자인 아이디어들을 가시화시킨다. 디자인 과정에서 오버레이 되고 다양하게 변하면서 만들어진 모형들 덕분에 아이디어가 발전하는 것이다. 기록하는 두 번째 방법은 여러 도시를 여행 다니면서 찍은 사진들을 디지털 파일로 보관하는 것이다. 잘 정리해두면 디자인 하는 동안에 무엇인가 찾아야 할 때 쉽게 찾을 수 있다.

DIMOS MOYSIADIS
XARIS TSITSIKAS

dmoysiadis.gr / x-t.gr

Dimos Moysiadis and Xaris Tsitsikas are collaborators since 2007. They run their own offices in Athens and Ioannina respectively with focus in the international architectural dialogue. They have established their own architectural language. Their approach towards architectural synthesis through an urbanistic perception is applied to strong and simple solutions that integrate with the city and result in novel spacial experiences. The core elements of their architectural philosophy consist in simplicity and clarity of the forms linked to the humane perception and scale.

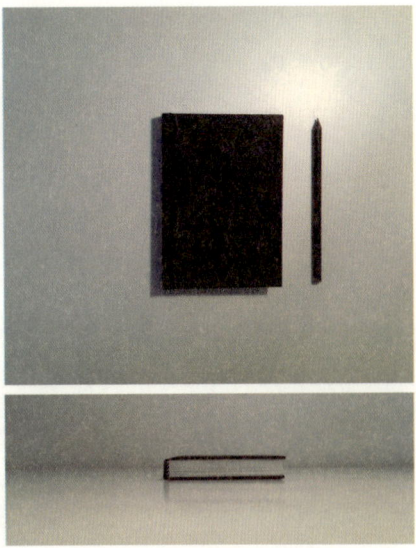

Interviewee: Xaris Tsitsikas(XT), Dimos Moysiadis(DM)

What is a notebook to you?

XT: It is a companion that allows you to be productive any time. The ideas are not only products of long hour focusing on a designing task. The incubation of an idea can awaken our enthusiasm on our day off or during lunch break. A sketchbook and a pencil are essential at that time.

DM: Sketchbook partly carries the notion of identity. There is a magical power that make everything drawn in a sketchbook more important than any of the delicate strokes on plain paper sheet.

당신에게 수첩이란 무엇인가?

XT: 수첩은 언제든지 작업을 할 수 있도록 도와주는 **동반자**다. 아이디어란 오랜 시간 동안 디자인을 해야지 나오는 것이 아니다. 잠자고 있던 아이디어는 쉬는 날이나 점심시간에도 우리의 열정을 일깨 울 수 있다. 이럴 때 스케치북과 연필은 꼭 필요한 것들이다.

DM: 스케치북은 자신만의 정체성을 지니고 있다. 그냥 흰 종이에 그려진 그 무엇보다 스케치북에 그려졌을 때 더 중요하게 느껴지도 록 만드는 신비한 힘을 가지고 있다.

Any episodes or memories related to a notebook?

DM: A particular booklet is the one that I have in my mind as my first sketchbook. It actually was. After the childish drawing blocks I was given by an architect, teacher of mine, a sketchbook and I was told to carry it everywhere. This was enough to be convinced. A strong bond is created between the one that holds the pencil and the sketchbook. I remember myself filling the pages with tremendous speed literally turning pages every minute. It was like a contest with no opponent. Today the pages are filled after deep thinking. Still fast but each of the strokes is meaningful. Even draft sketches of an architect reflects something more than the ideas represented.

수첩에 관련된 에피소드가 있다면 들려달라.

DM: 특별히 머릿속에 떠오르는 것은 나의 첫 스케치북이다. 건축가였던 나의 어렸을 적 선생님이 나에게 스케치북을 주면서 어디든지 항상 가지고 다니라고 했었다. 항상 가지고 다니라는 말 하나로 충분했다. 이로 인해 연필을 들고 있는 자와 그의 스케치북 사이에는 강한 연결고리가 생긴다. 그야말로 1분마다 한 장씩 넘겨가며 빠르게 그렸던 기억이 난다. 상대가 없는 대회를 하는 것 같았다. 요즘은 깊은 생각을 한 후에 페이지를 채워 나간다. 아직 빠르게 스케치를 하지만, 매 선마다 뜻이 담겨있다. 건축가의 스케치는 초안 조차도 보여지는 것 말고도 내제되어 있는 것이 있다.

When and where do you use your notebook the most?

It could be even in the car, if not while driving. It is a corny statement but you keep notes at any time since you woke up and you have not slept yet. Though where we both use our sketchbooks the most is in the offices while working.

수첩을 가장 많이 사용하는 공간과 때는?

운전하고 있지 않다면 차에서도 사용할 수도 있다. 좀 유치하게 들릴 수도 있지만, 일어나서 자기 전까지는 언제 어디서나 노트를 한다. 하지만 우리가 스케치북을 가장 많이 사용하는 시간은 **사무실에서 작업 할 때다.**

What influence does a notebook have in your projects and life as an architect?

XT: From my point of view the notebook has no influence in the way that the design proceeds. It might only affect the speed. Also there no importance if the cover is leather or dressed with fabric since the paper sheets work well with the pencil that you carry.

수첩은 당신의 작품과 삶에 어떤 영향을 끼치는가?

XT: 내가 생각했을 때는 수첩은 디자인 과정에 있어 어떠한 영향력도 없다. 그 과정 속도에는 영향을 미칠 수 있다. 그리고 표지가 가죽이든지 천이든지 상관이 없다. 가지고 다니는 연필과 잘 어울리는 것은 그 안에 든 종이이기 때문이다.

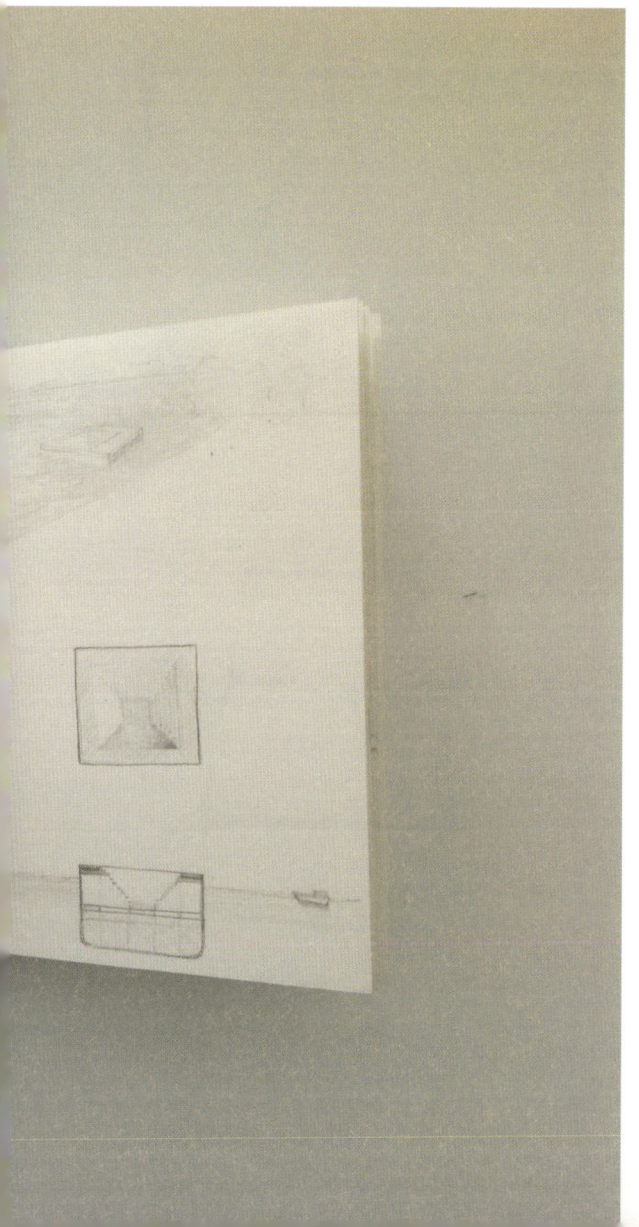

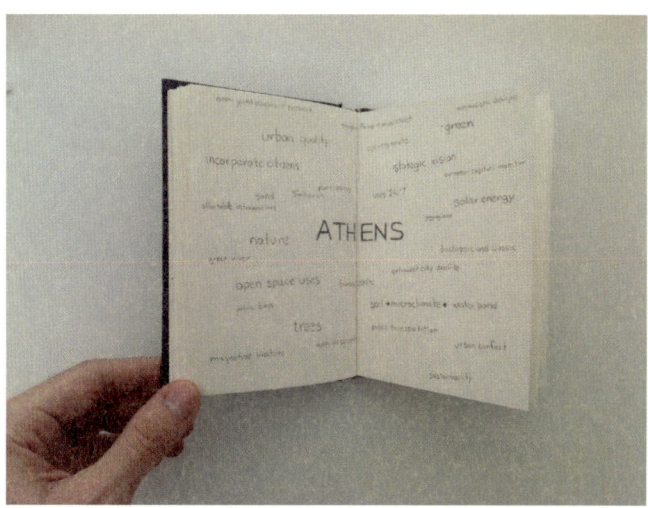

Are there anything else other than a notebook that you use to keep a record of your thoughts and ideas?

We both have spend tons of A4 format plane paper sheets. One sided sketches or two sided right before the recycle. It is been sometime that we keep our digital tablet sketchbooks. Not as handy but still it brings communication a step forward. Our working together for many years now make **the fancy sketches unnecessary** for the conception of an idea. The traditional sketchbook is more official in a way, though personal, the means **to convince oneself rather than a client,** a statement throughout time.

수첩 외에 자신의 생각을 기록하는 방법과 도구는 무엇이 있는가?

전에는 우리 둘 다 A4 용지를 가장 많이 사용했다. 한 쪽 면에만 스케치 하거나 재활용하여 양면으로 사용했었다. 하지만 요즘에는 태블릿을 사용한다. 스케치북만큼 편리하지는 않지만 소통하는 것에 있어서는 한 단계 올라섰다. 함께 일한 시간이 오래되어 이제는 아이디어를 전달하기 위해 **화려한 스케치는 필요 없어졌다.** 전통적인 스케치북은 개인적이면서도 어떻게 보면 좀 더 공식적인 면이 있기도 하다. 시간 속에 흘러가는 과정을 기록하며 **클라이언트보다는 나 자신을 확신시키는 것이다.**

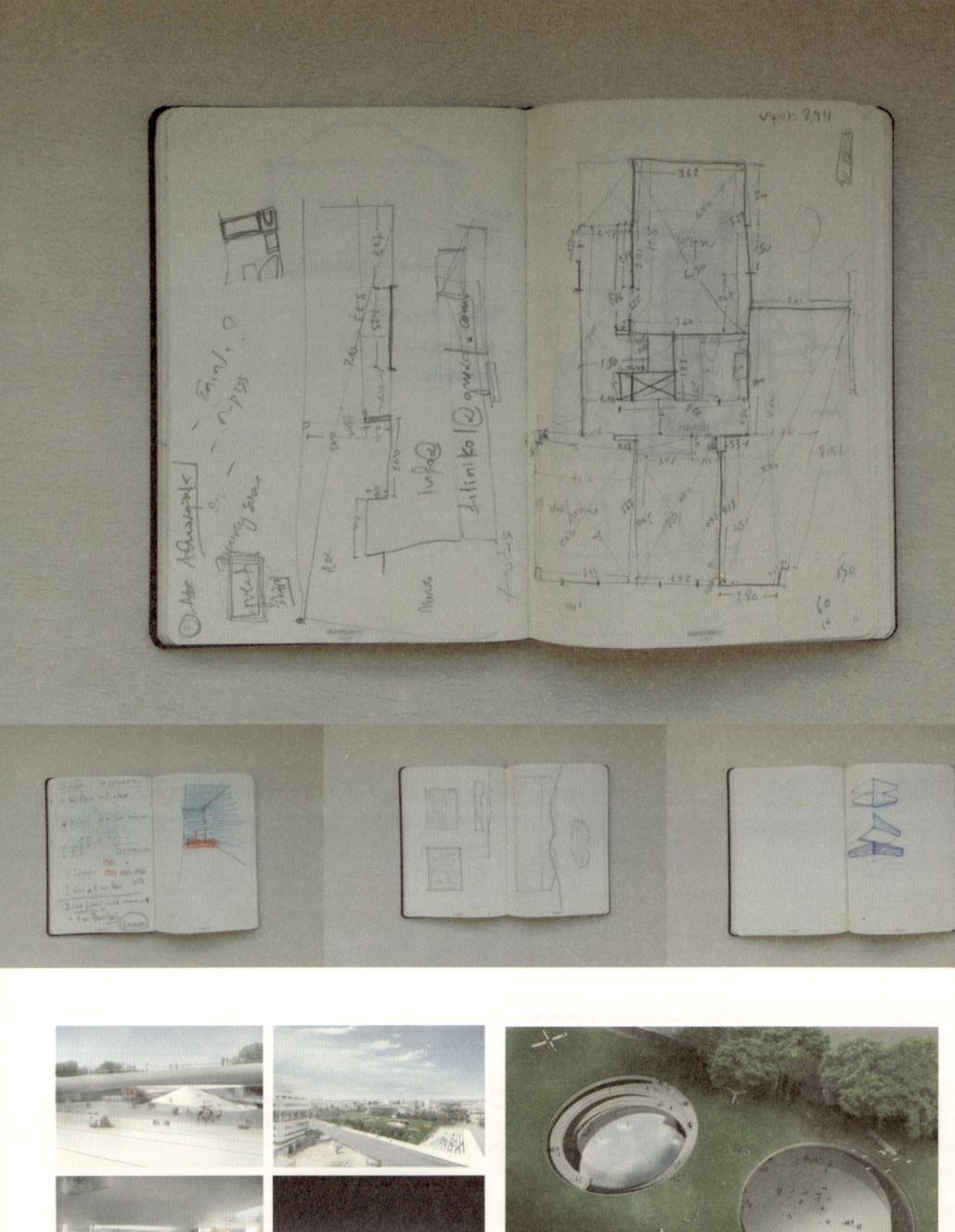

*ALEA

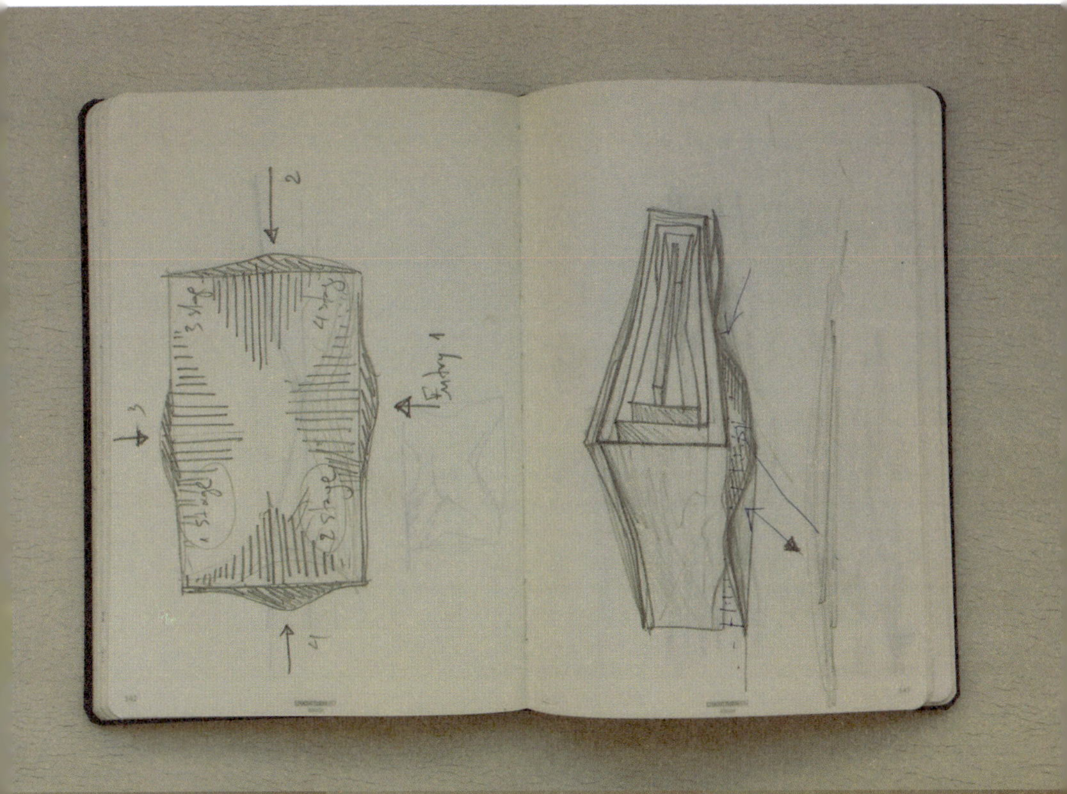

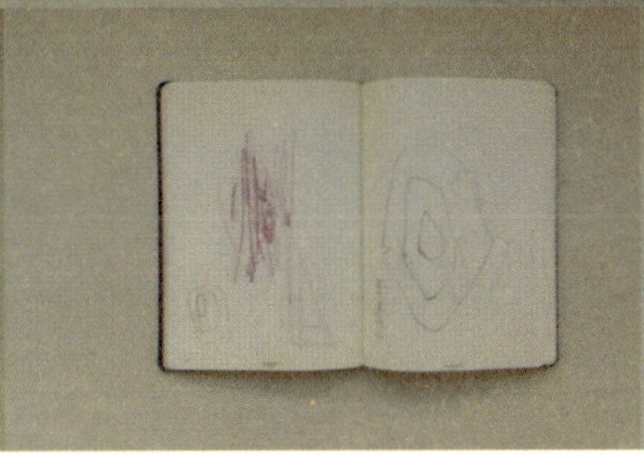

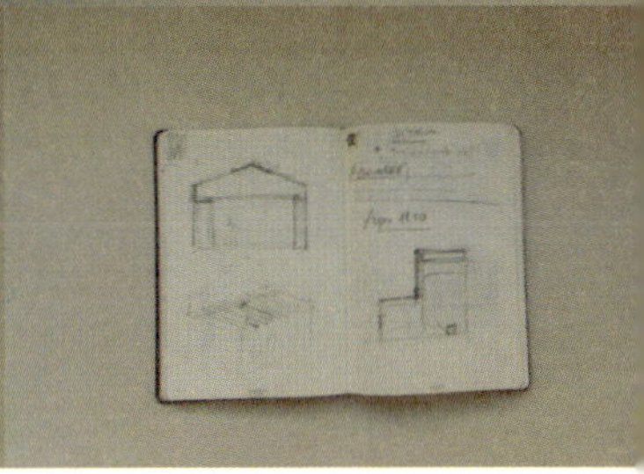

*BLEND

*EUREKA

EXTERNAL REFERENCE ARCHITECTS

www.externalreference.com

Awarded by New Italian Blood as the best Young Italian Architects of 2011, External Reference is an architectural office involved in design and research in the fields of interior design, architecture and landscape design.
Interested in generating intensity nodes the office is currently involved in research and design in the fields of exhibition spaces, galleries, hotels, accommodations and cultural events. Based in Barcelona the office establishes work collaborations with international professionals coming from different fields such as graphic design and art.
More of a professionals network rather than an office, External Reference is constantly questioning conventional configurations and solutions in order to engage, speculate and innovate.

External Reference notices that there is a big connection between evolvement of architecture projects and business models, and believes in sustainable development, balance between architecture, innovative strategy and business improvement.
Directed by Nacho Toribio and Carmelo Zappulla, External Reference offers a broad range of design strategies with the aim of pushing the boundaries towards new outcomes.
Looking for the unexpected the office team works on external factors such as client needs, budget limits and sustainability as driving forces for the design.
Developed and under construction projects have been developed in Spain, Korea, Italy, UK, Germany, France and Russia.

The architect's notebook is the object in which the thought process is captured with abstraction.

What is a notebook to you?

Undoubtedly, the notebook to the architect is the meeting point of human and hand-made aspects within a project. Within it, thinking and production develops in real time without technical intermediaries.
But, delving into what the pages of a notebook can offer, we think of it as blurred thinking territory, which surrounds the intellectual integrity of the architect. In the sketchbook not only do you find construction details or delicate perspectives but also thoughts, quotations, bibliographic references and an endless insight into the designers' life. Designers' sketches paradoxically merge with their daily lives (dates to remember, lists of materials for a model etc.). In this way, the end result when a book is completed, is close to a so-called 'artist's' book which makes it a medium of infinite possibilities: a time of freedom; being able to fill your pages with the representation of the passage of a professional life-time, the exceptional chance to go back in thought processes and ultimately, the possibility of reading spatiotemporal visual discourses of architecture. In short, **the architect's notebook is the object in which the thought process is captured with abstraction**, where it is at its most fully contingent, both in absolute size and in its micro manifestations expressed through rhythm and cadence.

당신에게 수첩이란 무엇인가?

건축가에게 수첩이란 사람과 프로젝트에서, 그 사람의 손길이 닿는 것들이 만나는 공간이다. 그 안에서는 어떠한 기술적인 요소를 없이 실시간으로 생각과 프로젝트가 발전한다.
하지만 수첩이 제공해줄 수 있는 것을 찾아봤을 때, 우리는 건축가의 지적인 진실성을 감싸는 희미한 생각의 공간으로 보게 된다. 수첩 안에는 도면 디테일이나 섬세하게 그린 투시도 스케치뿐만 아니라 생각, 인용, 참고 서적 그리고 끊임없이 디자이너의 삶을 통찰력 있게 바라보는 것들도 들어있다. 디자이너의 스케치는 기억해야 할 날짜, 모형 만들 재료와 같이 자신의 일상과 역설적으로 섞인다. 이렇게 스케치북을 다 썼을 때에는 무한대의 가능성을 보여주는 일명 '아티스트의 책' 이 된다. 스케치북 안에 전문가로서의 자신의 삶을 표현할 수 있고, 지나온 생각 과정을 되돌아 볼 수 있는 특권이 주어지며, 시공간을 넘나드는 건축을 시각적으로 읽을 수 있는 수첩은 자유로운 시간이 된다.
다시 말해, **건축가의 수첩은 생각 과정을 추상적으로 담는 물건이다.** 리듬과 억양으로 표현된 아주 소소한 표시나 크기를 임시로 담는다.

Any episodes or memories related to a notebook?

Interestingly, the architect's notebook, read after some time, always produces a certain tenderness in the sense that within its pages, the designer relates to their surroundings with more freedom, open-minded thinking, fearlessness and innocence. In this way, it becomes a kind of dialogue of censored options that have never been subjected to materiality, or the monstrosities designed without passion, that end up discarding the concrete reality of construction, economy and program.
Therefore our notebooks give off a certain rebelliousness and **we discover the testimony of how through dreaming, in every moment, the spirit of each project is generated.** I have come to realise over time, one can reach a maturity akin to the original sketch, although in many cases surprisingly far in form and materiality, it remains substantially tangent to its intellectual genetics.
No doubt, if we were to tell the story of the office, we would dust off each sketchbook, and find individual reflection places, representing individual concerns within collective Professional consensus. And at this point, there are exceptional events as the drawings demonstrate a time where the hand of a member of the practice is at the service of the thoughts of another. A true professional emotional time.

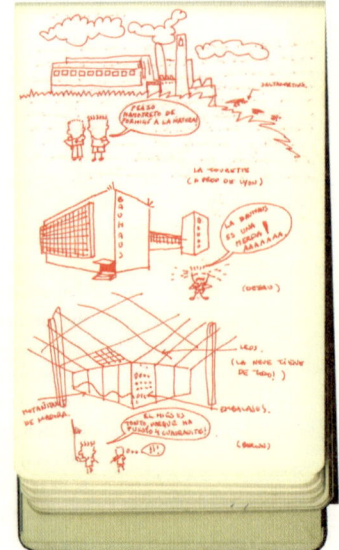

수첩에 관련된 에피소드가 있다면 들려달라.

신기하게도 건축가의 수첩은 시간이 지난 후 다시 들여다보면, 그 페이지 속에서 특정한 감정이 떠오른다. 디자이너는 좀 더 자유롭게 자신의 주변을 이해하고 두려움 없이, 천진함과 오픈 마인드를 가지고 생각한다. 이러했을 때 물리적인 것이나 시공, 경제, 그리고 프로그램의 명확한 현실을 거부하게 되는 열정이 없는 흉물스러운 디자인에 얽매이지 않은 가능성들을 나눌 수 있다.
그러므로 스케치북은 무언가 반항적인 느낌을 주고, **우리는 프로젝트의 혼이 매 순간 꾸는 꿈을 통해 어떻게 생겨나는지를 알게 된다.** 나는 시간이 지나면서 초기 스케치의 형태와 재료는 많이 다르지만 지성적인 요소와 그의 성숙함은 비슷하게 유지된다는 것을 깨달았다. 만약 사무실에서 이 얘기를 들려주고 있다면 우리는 아마 각 스케치북의 쌓인 먼지를 털어 추억이 담긴 곳을 하나하나 보여줄 것이다. 그리고 요즘에는 이 스케치를 그린 사람의 손이 지금은 다른 사람의 생각을 표현하고 있는 일들이 있어, 감성적인 시간들이다.

We discover the testimony of how through dreaming, in every moment, the spirit of each project is generated.

When and where do you use your notebook the most?

Perhaps to concretise the notebook in a specific time and place is the opposite of its genuine definition. One of the hallmarks of professional and personal identity of an architect is to be always accompanied by his notebook. Thus, the almost existential need of an architect drawing can not be separated from the notebook. Thus, the place and time to use the drawing does not exist. Any time is right to reflect drawing.
Even while sleeping; **how many times does a designer wake up in the middle of the night to draw sketches, often incomprehensible, generated in their night dreamlike drifts.** Sometimes even when one sketched on a napkin in a bar or on a wall with a piece of chalk, sketches become precarious. Undoubtedly, the architects notebook is full of these napkins, nurturing an even greater vitality.

What influence does a notebook have in your projects and life as an architect?

Using a notebook everyday as a professional tool, inevitably offers a constant critical attitude to our projects that never inhabit the world of digital processes.
Moreover, the notebook offers the project a delicacy of reflection in terms of materiality and construction processes, as freehand drawing is a process of physical and intellectual construction itself. With all this, a sketchbook gives the architect the ability to stay in touch with the traditional tools of the profession, which capture the most humanistic aspects of the project.

수첩을 가장 많이 사용하는 공간과 때는?

수첩을 특정한 시간과 장소로 구체화 시키는 것은 수첩의 본 뜻과 반대되는 것이 아닌가 생각된다. 건축가가 자신의 전문성과 정체성을 나타내는 특징은 항상 수첩을 지니고 다니는 것이다. 그러므로 건축가의 도면이 필요한 것과 수첩을 따로 생각할 수 없고, 그림을 그리는 시간과 장소 또한 정해져 있지 않다. 언제나 그림을 그릴 수 있기 때문이다.
잠을 자다가도 꿈을 꾸다 떠오른, 주로 이해할 수 없는 아이디어들을 스케치 하느라, 한밤에 일어나는 횟수가 얼마나 될까? 냅킨에 스케치한 것이나, 분필로 벽에 스케치 한 것이나, 스케치들이 불확실할 때도 있다. 하지만 확실히, 건축가의 수첩 안에는 수두룩한 이러한 냅킨들이 하나의 활력소가 된다.

수첩은 당신의 작품과 삶에 어떤 영향을 끼치는가?

전문적인 도구로 수첩을 매일 사용하면, 디지털 세계에서만 진행되는 프로젝트를 끊임없이 분석적인 사고로 바라볼 수 있도록 해준다. 또한, 수첩은 프로젝트를 재료성과 시공 과정을 보다 더 섬세하게 들여다 볼 수 있도록 해준다. 프리핸드 스케치하는 그 과정 자체도 물리적이고 지성적으로 구성 되어 있기 때문이다. 더해서 스케치북은 건축가에게 프로젝트의 가장 인간적인 면인 전통적인 도구들을 계속 사용할 수 있도록 해준다.

Sant pere de ribes, Spain

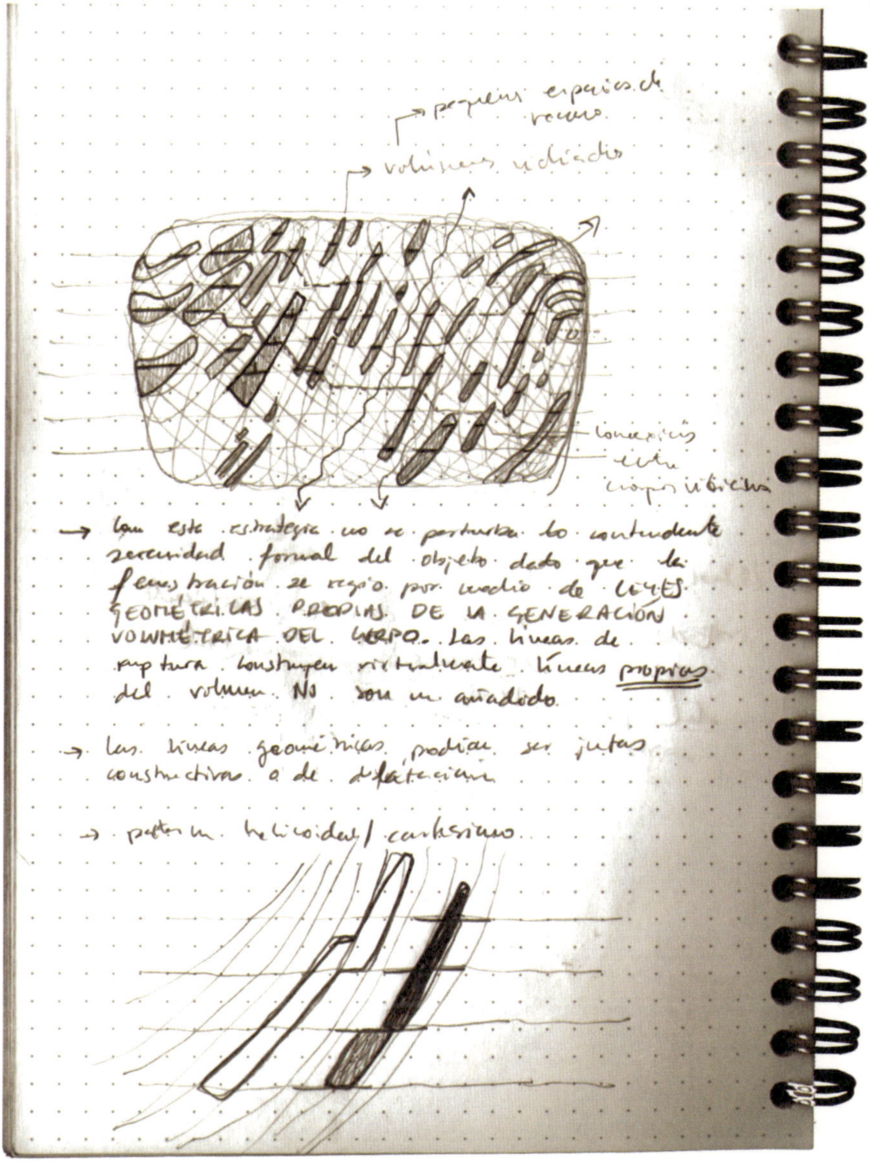

Juzgados d elo penal, Spain

→ pequeños espacios de vacío
→ volúmenes reductores

→ Con esta estrategia no se perturba lo contundente y rotundidad formal del objeto dado que la fenestración se regia por medio de LEYES GEOMETRICAS PROPIAS DE LA GENERACIÓN VOLUMÉTRICA DEL CUERPO. Las líneas de ruptura constituyen virtualmente líneas propias del volumen. No son un añadido.

→ Las líneas geométricas podrían ser juntas constructivas o de dilatación.

→ patrón helicoidal cartesiano.

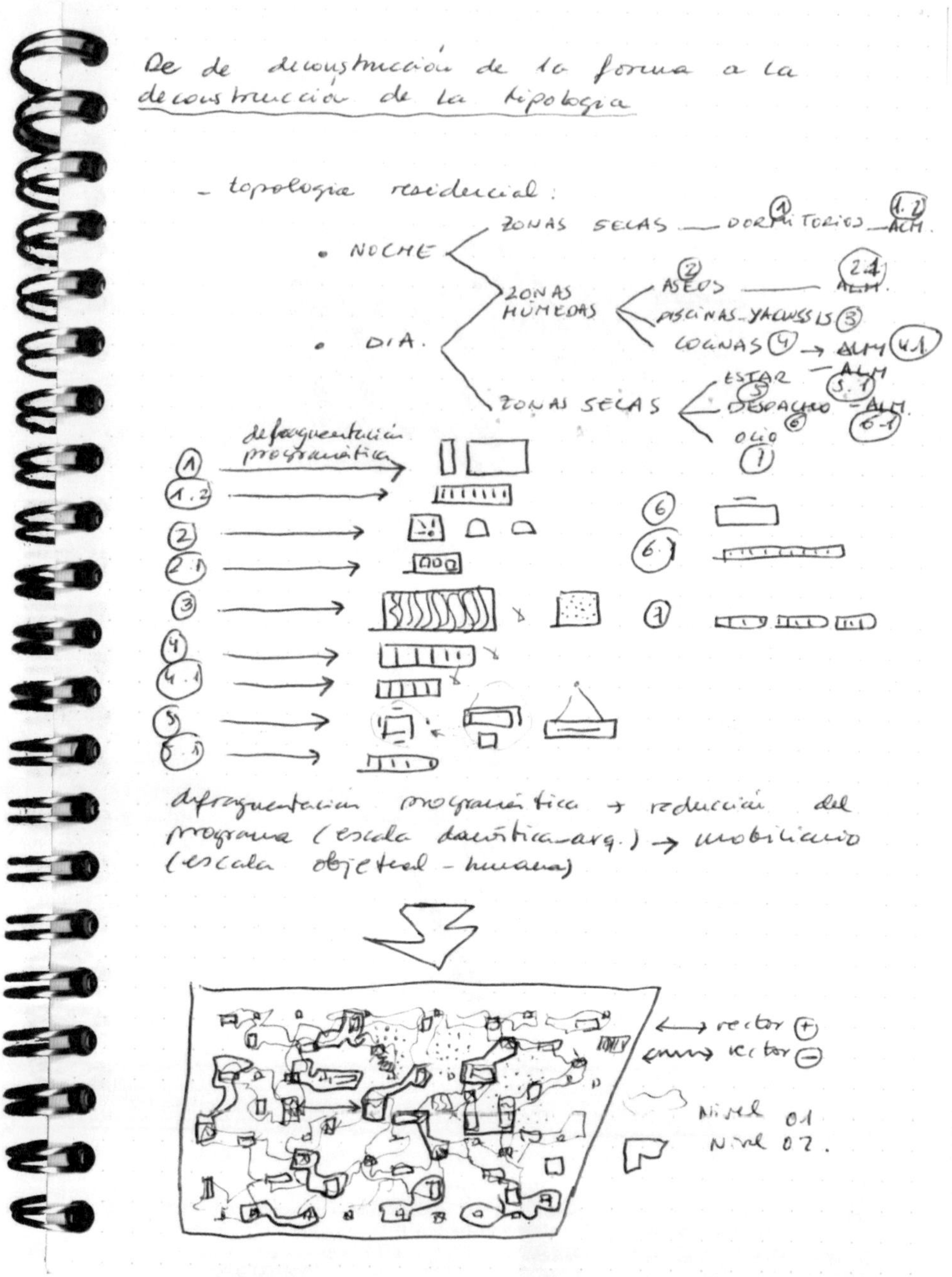

***Jarvenpaa**, Finland*

De de deconstrucción de la forma a la
deconstrucción de la tipología

- tipología residencial:

  • NOCHE ⟨ ZONAS SECAS — DORMITORIOS ① 1.1 ALM

  ⟨ ZONAS HÚMEDAS ⟨ ASEOS ② — 2.1 ALM
                   PISCINAS-JACUSSIS ③
                   COCINAS ④ → ALM 4.1
  • DIA.   ESTAR ⑤ — ALM 5.1
  ⟨ ZONAS SECAS ⟨ DESPACHO ⑥ — ALM 6.1
                   OCIO ⑦

desfragmentación programática →

① →
1.2 →
② →
2.1 →
③ →
④ →
4.1 →
⑤ →
5.1 →

⑥
6.1
⑦

desfragmentación programática → reducción del
programa (escala doméstica-arq.) → mobiliario
(escala objetual-humana)

⇩

⟷ rector ⊕
⟷ rector ⊖

Nivel 01
Nivel 02

Public Services
01_workshop school
02_public sauna
03_arts and crafts museum
04_health center
05_house for elders
06_exhibition space
07_library
08_sport center
09_kindergarden
10_playground
11_restaurants
12_night bars
13_social club
14_neighbours association
15_gym
Canals
Vegetation
Main roads
Bus
Bike paths
Pedestrian paths
0 10 50 100

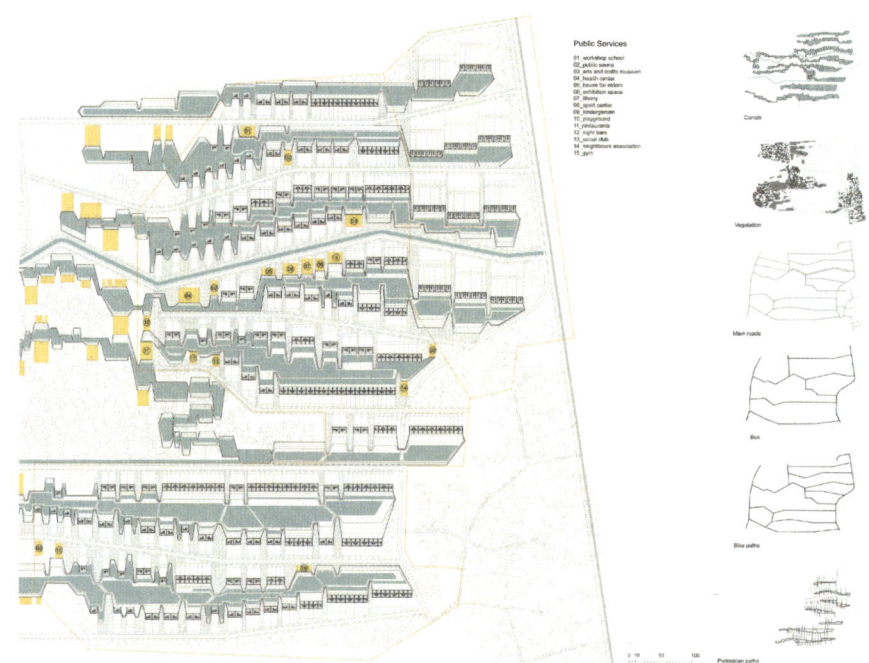

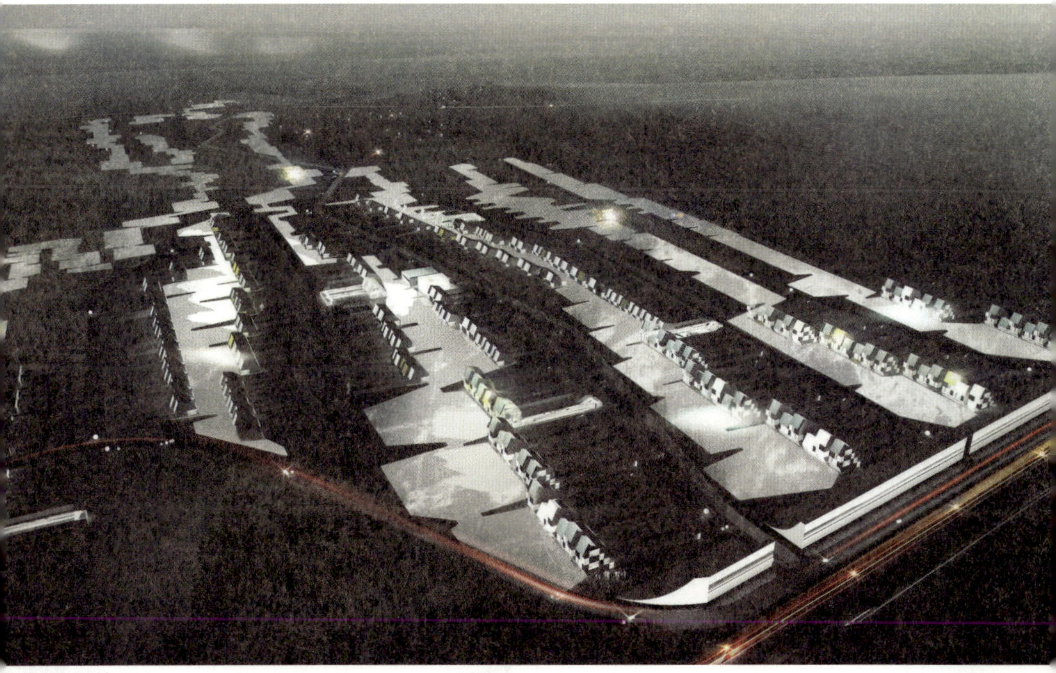

Are there anything else other than a notebook that you use to keep a record of your thoughts and ideas?

In addition to the nostalgic enrichment of the notebook, the immediacy of communication that today's technology offers is invaluable. On one hand, through smartphone technology, communication between members of the office have almost surpassed the format of e-mail communication to impose ´chat´ applications. Thus, the smartphone allows us to share images and scenarios when practice members are not together. **Our iPhone has almost become a sort of 'collective eye' or 'common memory'.** When we're apart and when there is a reference that interests us, architectural or not, or when work problems arise, a photograph sent by chat and a brief dialogue which is often only understandable by ourselves, become an essential professional communicative mechanism. So, the message history, photographs and thoughts in our messaging thread, is a way of consolidating a kind of common digital sketchbook that we can access by sliding our finger across the touchscreen, just as we turn, the yellowish pages of our old notebooks.

수첩 외에 자신의 생각을 기록하는 방법과 도구는 무엇이 있는가?

향수를 불러일으키는 수첩 말고도, 오늘날의 기술이 제공해 주는 신속한 커뮤니케이션 또한 소중하다. 스마트폰 기술을 통해 사무실 사람들과 소통하는 방법으로 이메일이 아닌 '채팅' 앱을 사용하게 되었다. 그래서 사무실에 같이 있지 않은 사람들에게도 이미지나 시나리오를 나눌 수 있도록 해준다. **아이폰은 거의 우리 모두의 '눈' 또는 '기억'이 되었다.** 우리가 서로 떨어져 있는 상황에서 참고하기 좋은 내용이 있을 때나, 하는 일에 문제가 생겼을 때에, 채팅 창으로 보내는 사진과 우리밖에 이해할 수 없는 짧은 대화가 전문적으로 소통하는 방법으로도 매우 중요한 부분을 차지한다. 그래서 메시지 히스토리나 사진들, 그리고 대화 속에 적혀있는 생각들이 하나의 디지털 스케치북이 된다. 낡은 스케치북의 한 장 한 장을 넘기듯, 손가락으로 화면을 넘긴다.

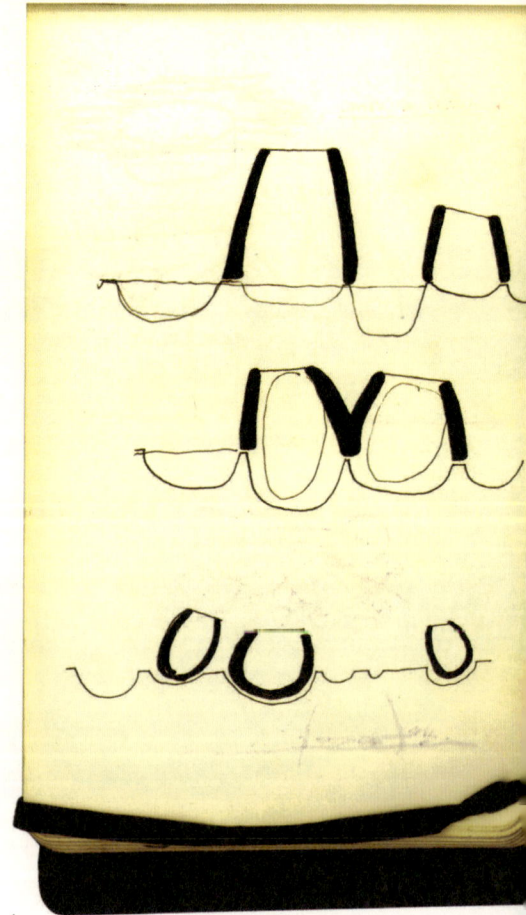

***Vespella de gaià**, Spain*

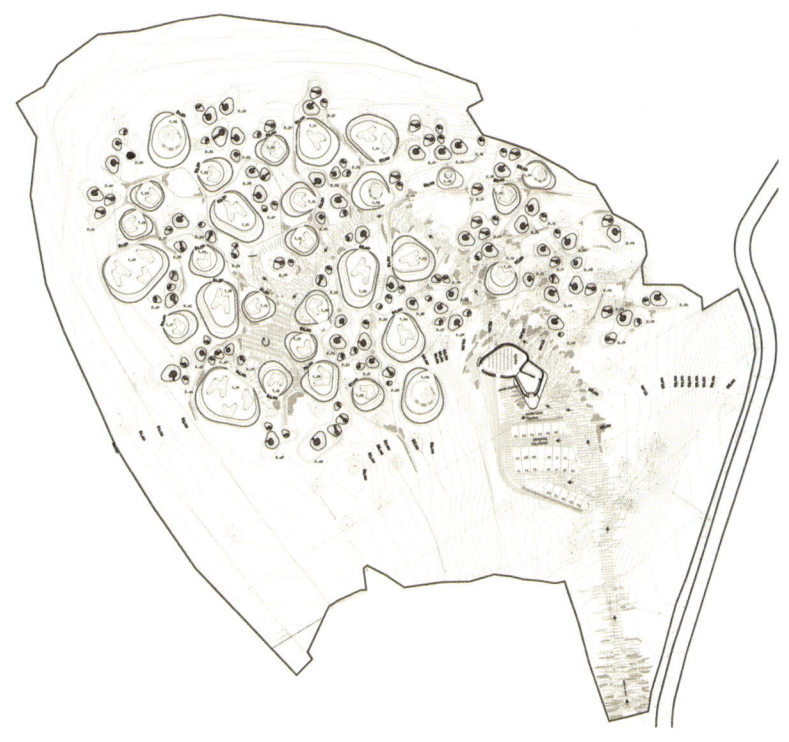

*Enjoy barcelona_reception desk

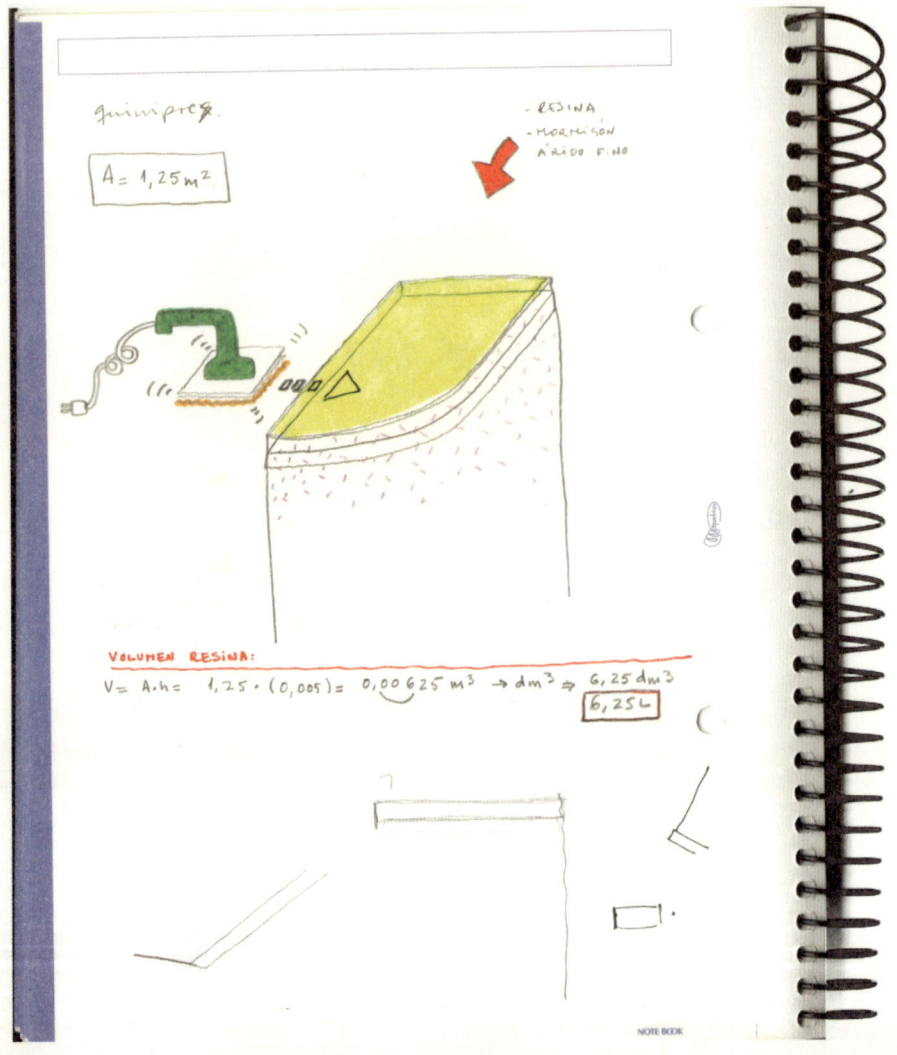

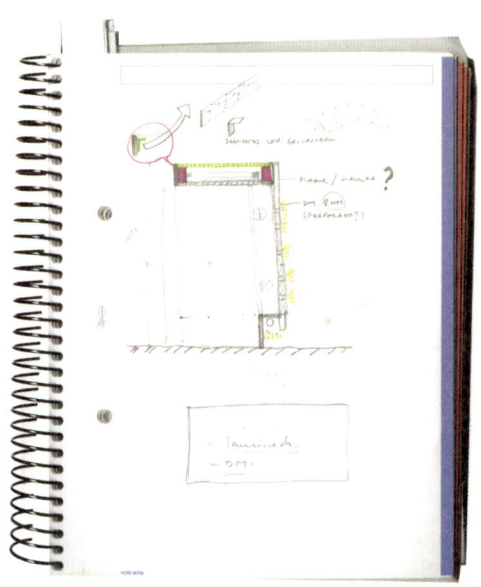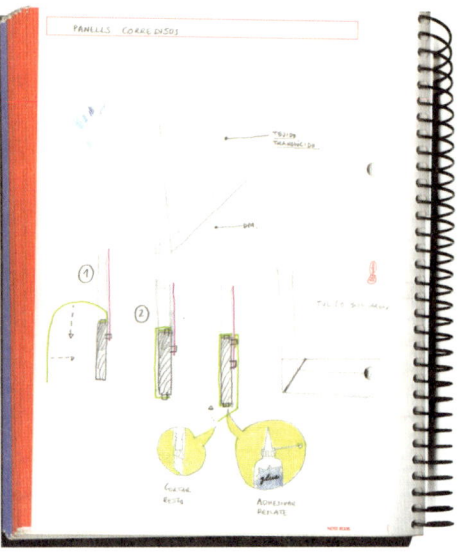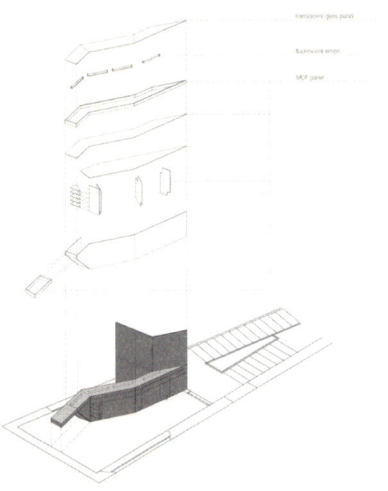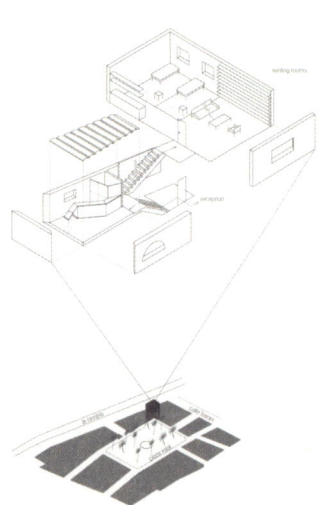

Irkush skyscraper, Russia

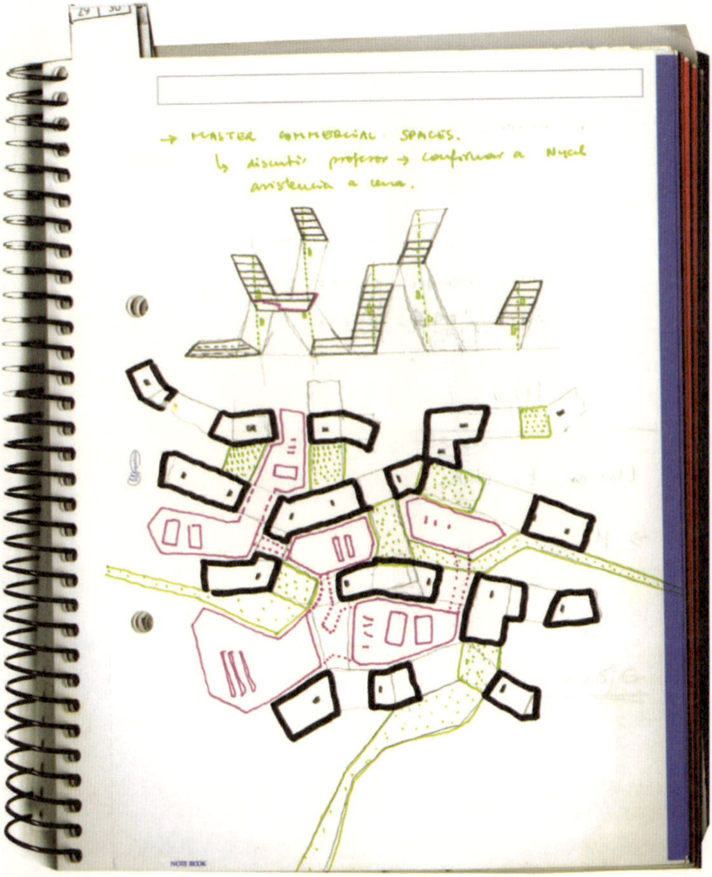

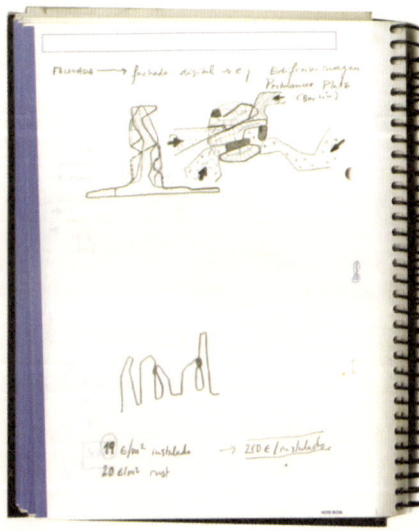
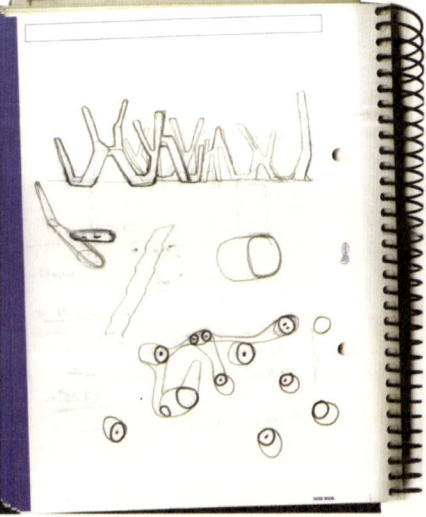

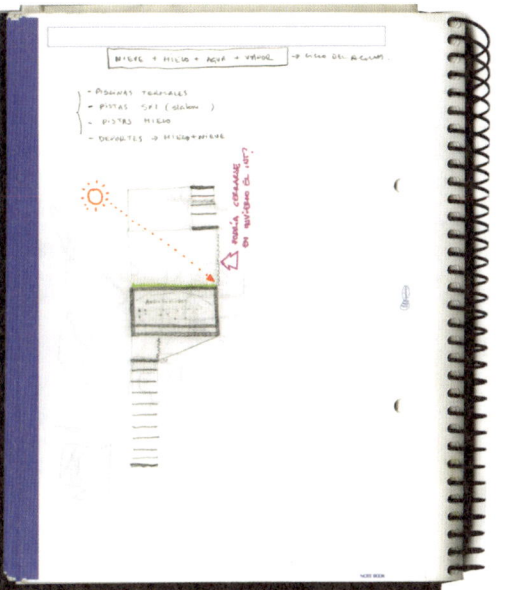
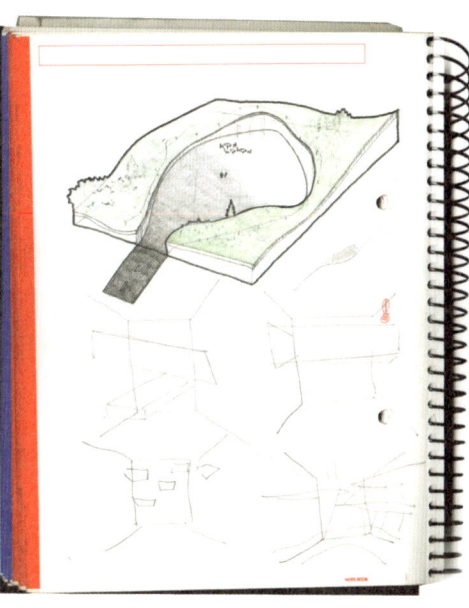
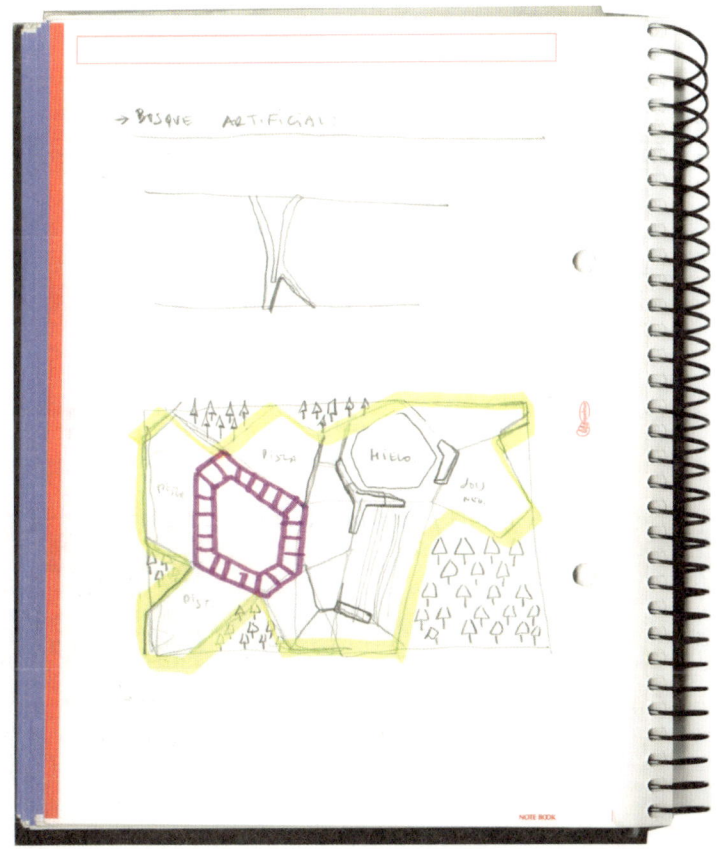

*RAS gallery

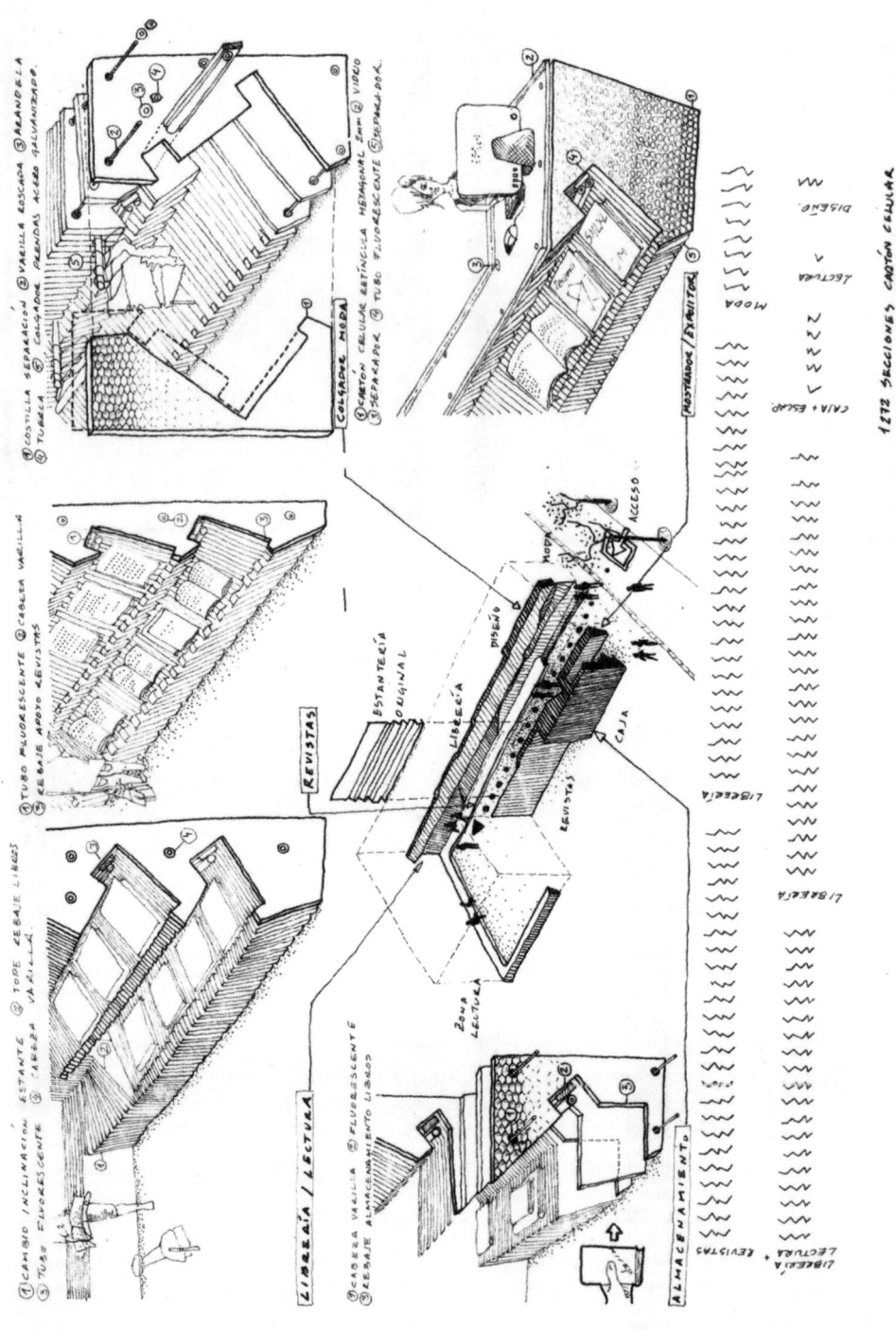

Hotel juan carlos I

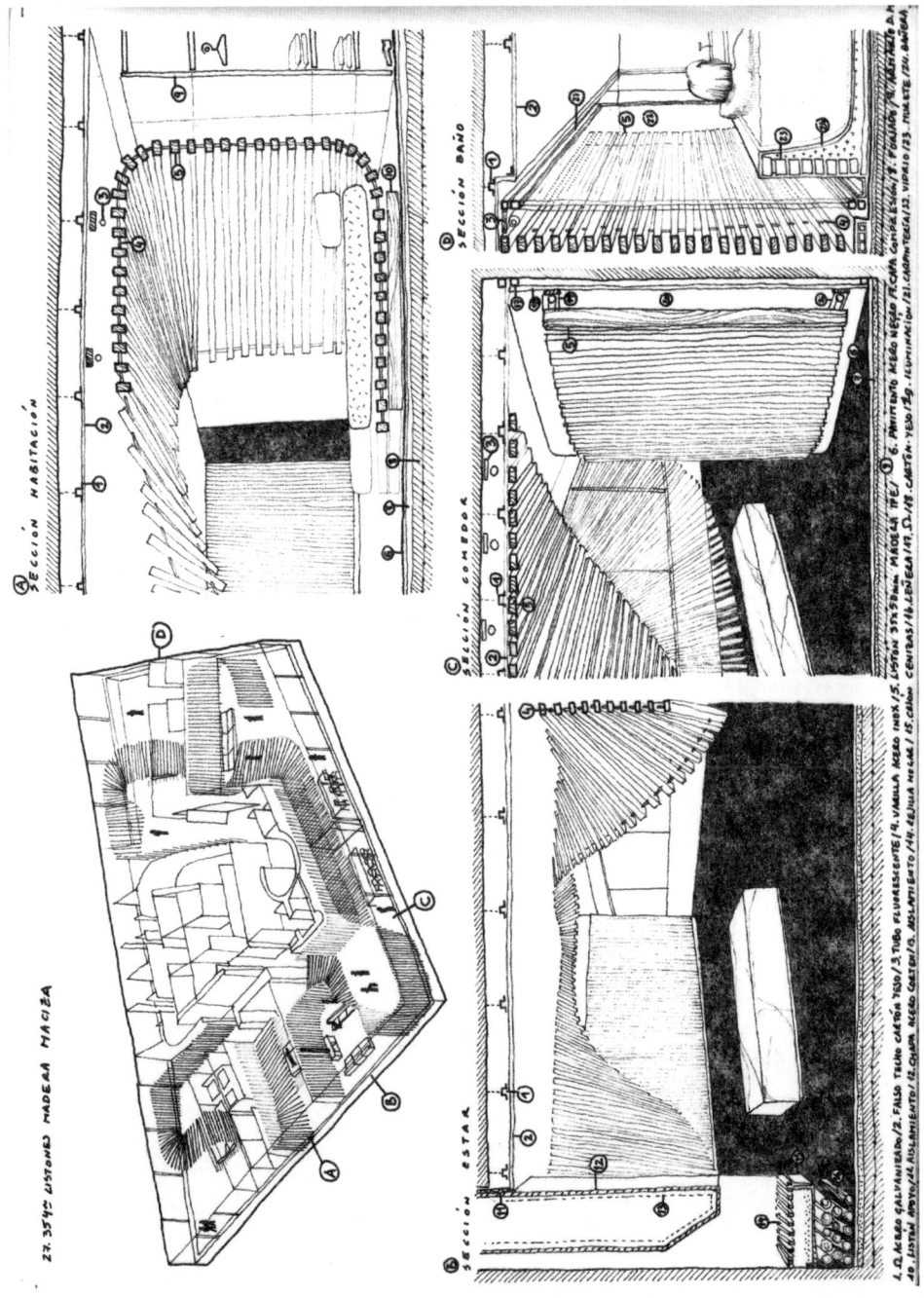

OOIIO Architecture

www.ooiio.com

OOIIO Architecture is an international design team focused on singular personalized solutions for architectural needs. We start every project from a deep understanding of the context and the client requirements and we provide innovative, contemporary and unique architectural answers to satisfy them.
OOIIO´s team was educated, and also has been teacher, of some of the most prestigious Architecture Universities of the world. Before the studio´s foundation they have been working together with some of the most influential and revolutionary architects of last decades, like Sir Norman Foster or Rem Koolhaas, both winners of Architecture Pritzker Prize (Nobel equivalent), where they participated and had senior responsibilities in several designs and constructions for buildings all around the world.
Some years ago they decided put into practice all this gain knowledge and founded OOIIO, offering a first international class design service for a highly competitive prize, appropriate for the different markets reality.
OOIIO has its headquarters in Madrid, Spain.

What is a notebook to you? 당신에게 수첩이란 무엇인가?

I do not have a notebook; I collect papers, models, ideas, etc. in boxes. **For me those boxes keep the project soul.**

My notebook is any white piece of paper or board that I can write, draw, cut, etc. I do not have a physical notebook, my notebook is a mountain of papers that are always all around me... included little sketch models that I produce constantly (I do not like work with the computer on 3D). I prefer to model with my hands. It gives always a "human touch" to every project.

This big collection of random documents is my notebook, which I collect like a little treasure, because it shows all the project process: first ideas, shape and program distribution evolution, rejected options,... and also short sentences that I wrote while thinking on the building showing he worries I have during the design process, that always influence on the final result, like "let´s push harder", "I need to call to this guy", "I need to pay this bill", "do not have time", "at 18:00 h I will meet that guy"...etc

A notebook is a piece of your live, and it is made of lots of different things, that we keep in boxes.

나는 수첩을 가지고 있지 않다. 종이나 모형, 아이디어들을 적은 것들은 상자에 보관한다. **프로젝트의 영혼이 담긴 상자가 된다.**

쓰고, 그리고, 자를 수 있는 모든 하얀 종이나 판이 나의 수첩이 된다. 나는 실질적인 수첩을 가지고 있지 않고, 내 주변에 산처럼 쌓인 종이들이 나의 수첩이 된다. 이 중에는 틈틈이 만드는 스케치 모델들도 포함되어 있다. 나는 컴퓨터로 3D 작업을 하는 것보다 직접 손으로 만드는 것을 좋아한다. 프로젝트마다 '사람의 흔적'을 남기기 때문이다.

내가 보물같이 모으는 이 무작위한 종이들이 바로 나의 수첩이다. 첫 아이디어부터 형태와 프로그램의 진화 과정부터 버려진 선물까지, 프로젝트의 모든 과정을 보여주기 때문이다. '조금만 더 힘을내자', '이 사람한테 전화해야 함', '청구서 낼 것', '시간이 없다', '오후 6시에 저 사람과 만남' 등, 디자인하면서 고민하고 생각하며 썼던 짧지만 최종 디자인에 영향을 끼치기도 하는 문장들도 포함되어 있다.

수첩은 상자 안에 담아두는, 여러 가지로 만들어진 삶의 한 조각이다.

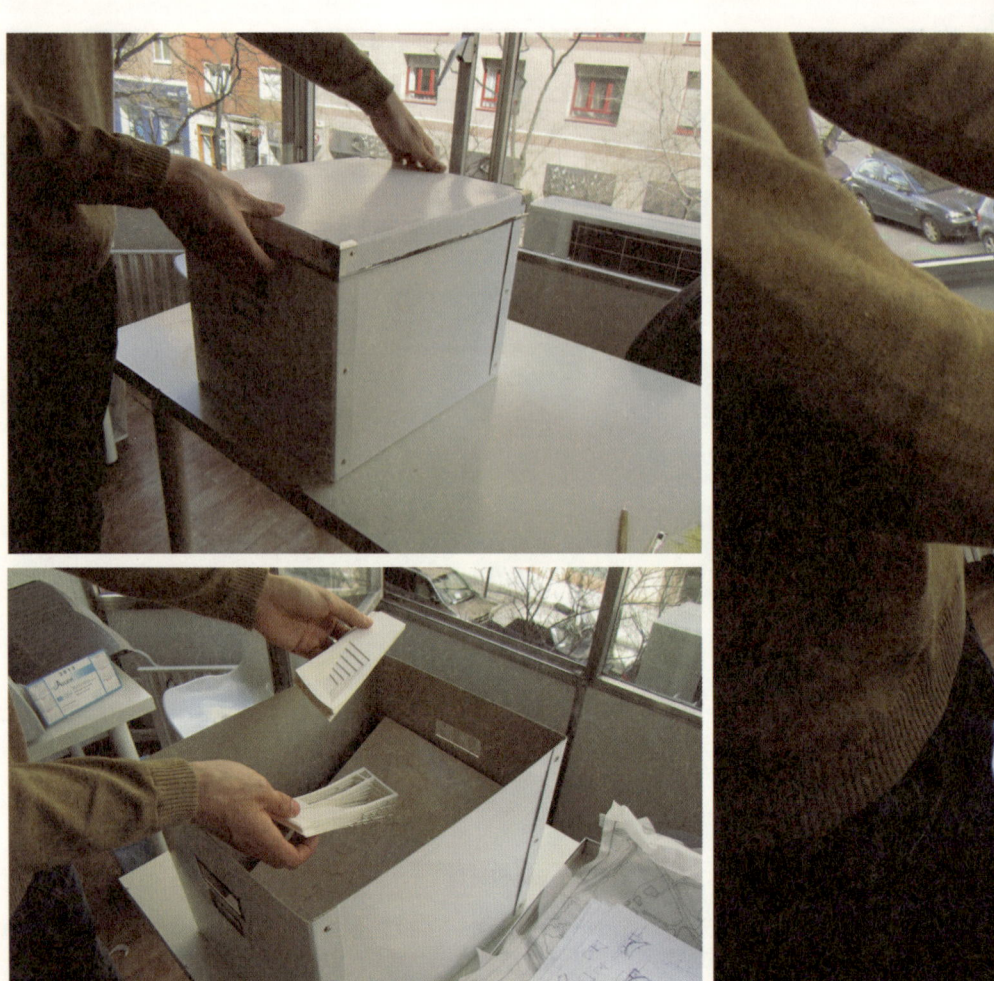

Any episodes or memories related to a notebook?

We use any kind of paper to draw in, and usually advertising papers that we recycle, or wasted printouts from other projects, etc.
Go at the end of the day, when you see the box-sketchbook for a particular project, **you discover that we have used the test printouts for that other project that we were designing at the same time on our office, and suddenly lots of remembers of that time come to your mind, people that was working on the team, old project options, etc.** and the most incredible think is that you discover how the projects done at the same time on the office have some design connections, that you didn´t realize at that time. But now you can see watching those old documents.

수첩에 관련된 에피소드가 있다면 들려달라.

우리는 그릴 수 있는 종이라면 아무거나 쓰는데, 대부분 광고지나 다른 프로젝트에서 나온 이면지들을 재활용한다. 그래서 특정 프로젝트의 스케치북 상자를 들여다보면, **이면지로 사용된 같은 시기에 작업했던 다른 프로젝트의 도면들이 보인다. 그 도면들을 보면 같이 작업했던 팀 사람들, 예전 옵션들 등, 그 프로젝트에 대한 추억들이 떠오른다.** 이렇게 생각을 하다 보면 같은 시기에 작업했던 프로젝트 디자인들이 서로 관련되어 있다는 것이 보이기 시작 한다. 그 당시에는 몰랐지만, 이렇게 예전 도면이나 그림들을 보면서 알게 되는 것들이다.

When and where do you use your notebook the most?

I try to use the sketchbook only at my studio or when I am working outside (client presentation, construction site, when I am on a lecture or exhibition, etc) I try to NEVER think on work when I am on holidays or with friends, or so. Notebooks are forbidden then!

수첩을 가장 많이 사용하는 공간과 때는?

스튜디오에 있을 때나 밖에서 일할 때만 (건축주 프레젠테이션, 현장 방문, 강의나 전시) 수첩을 사용한다. 휴가를 갈 때나 친구들과 있을 때에는 절대 일에 관련된 생각을 하지 않으려고 한다. 이럴 때에 수첩은 금지!

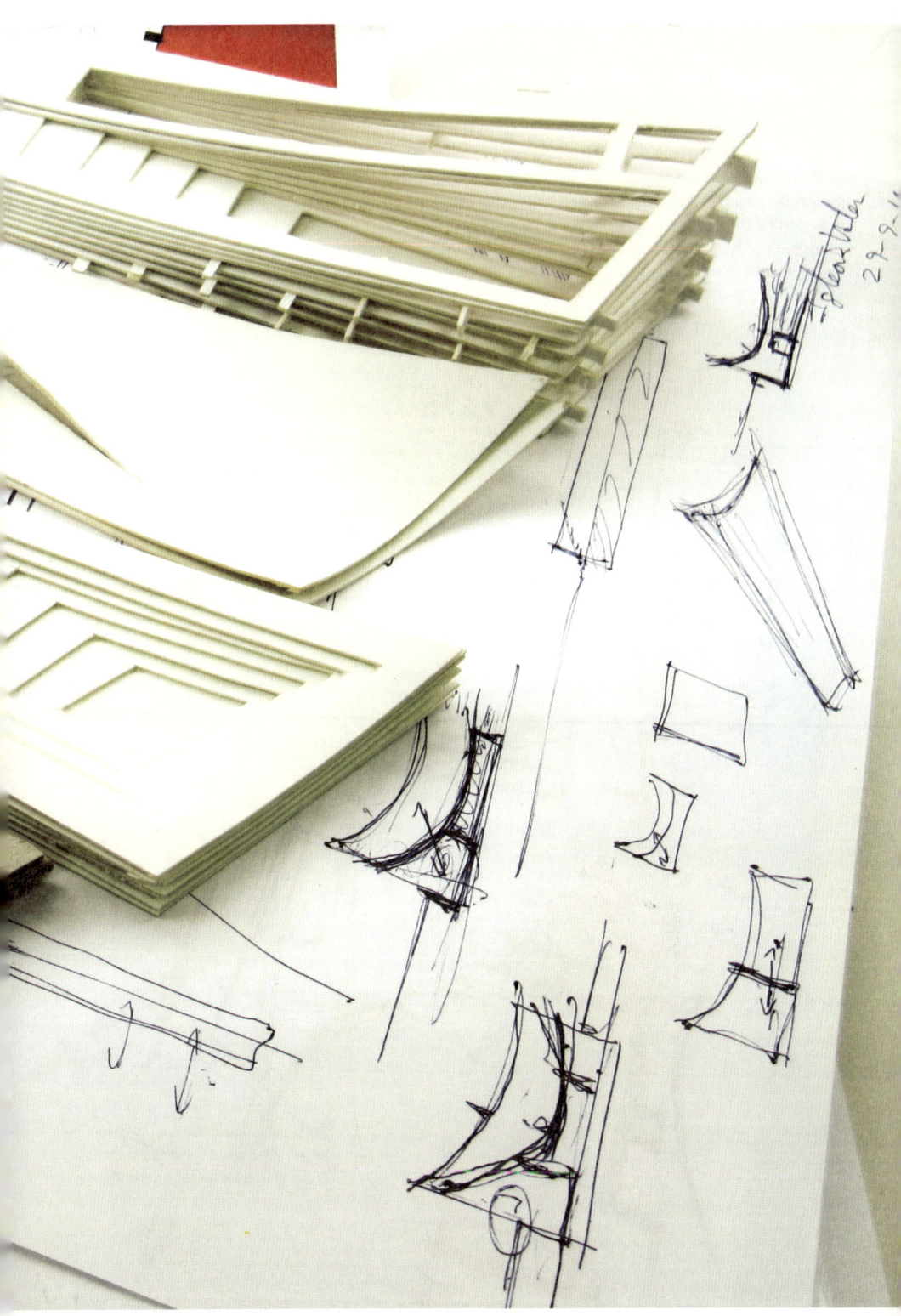

What influence does a notebook have in your projects and life as an architect?

Our box-sketchbooks are the real project.
For me the project is not the final result, this is just a moment of the project. Actually the building is what happens since the first client meeting and the today's day user that is getting in or using it... and the box keeps the soul of how the building was born.

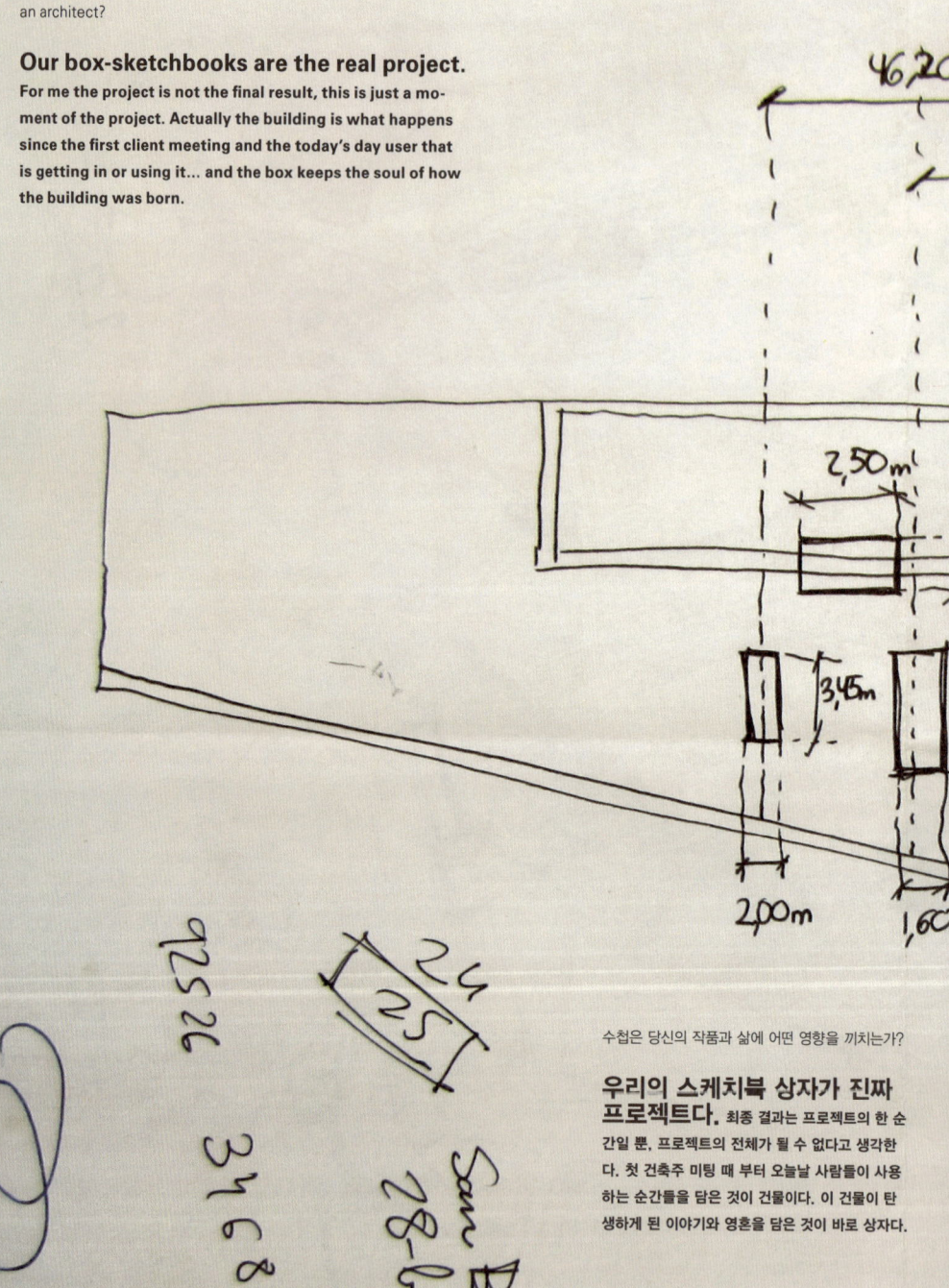

수첩은 당신의 작품과 삶에 어떤 영향을 끼치는가?

우리의 스케치북 상자가 진짜 프로젝트다. 최종 결과는 프로젝트의 한 순간일 뿐, 프로젝트의 전체가 될 수 없다고 생각한다. 첫 건축주 미팅 때 부터 오늘날 사람들이 사용하는 순간들을 담은 것이 건물이다. 이 건물이 탄생하게 된 이야기와 영혼을 담은 것이 바로 상자다.

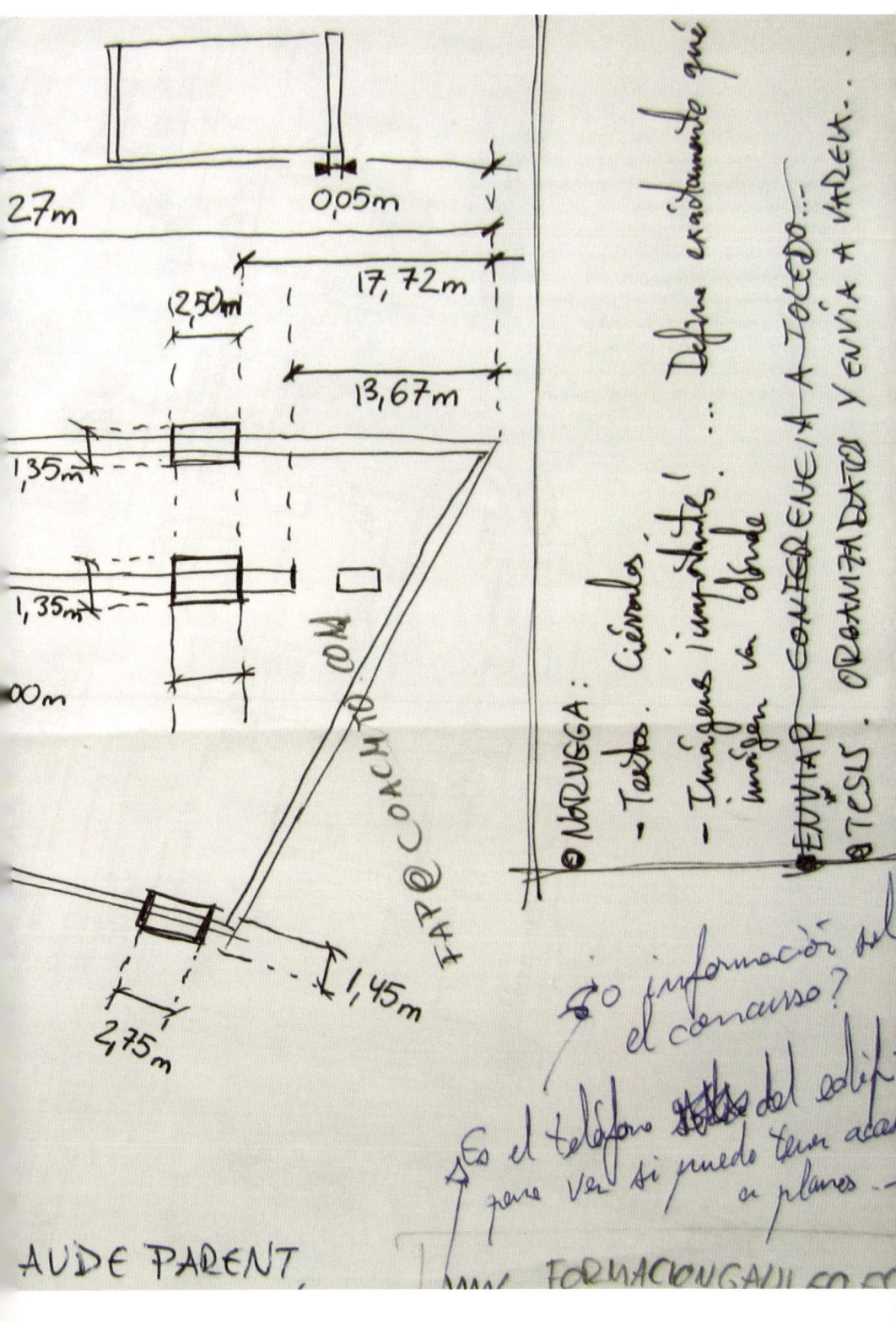

Are there anything else other than a notebook that you use to keep a record of your thoughts and ideas?

As I said, we do not have traditional notebooks on our studio. We collect every kind of notes: at one side we stack papers and papers with ideas, printouts, data, etc. on the other side 3D sketches made randomly with cheap materials for our volumetric studies, on the wall in front of us we hang the selected ideas and they become like "project design guides" o "design directions".... And at the end, when we finalize our work we collect everything on boxes, that we keep as little treasures.
(always in every document that we write, we put the date and the project, for the final collection).

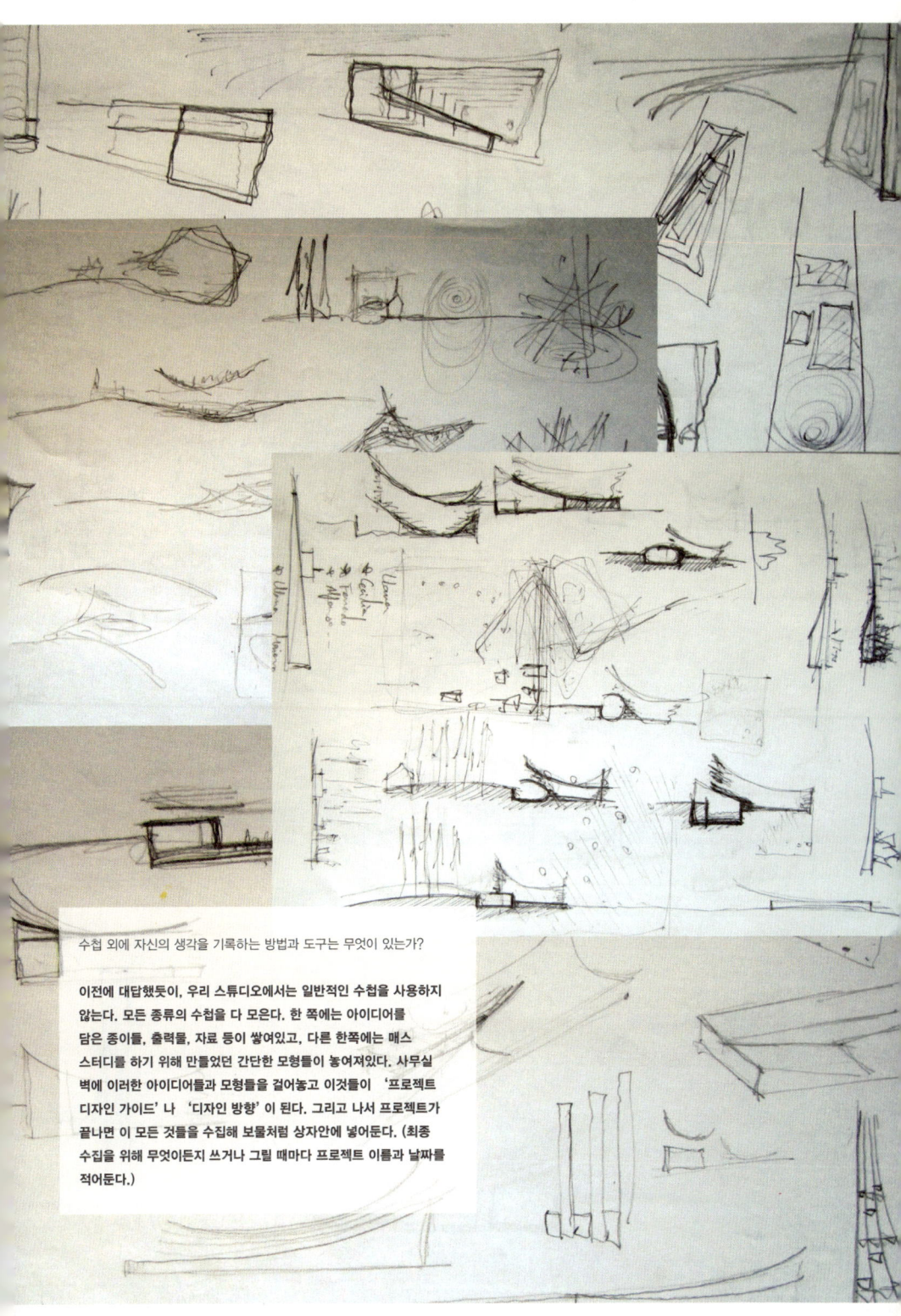

수첩 외에 자신의 생각을 기록하는 방법과 도구는 무엇이 있는가?

이전에 대답했듯이, 우리 스튜디오에서는 일반적인 수첩을 사용하지 않는다. 모든 종류의 수첩을 다 모은다. 한 쪽에는 아이디어를 담은 종이들, 출력물, 자료 등이 쌓여있고, 다른 한쪽에는 매스 스터디를 하기 위해 만들었던 간단한 모형들이 놓여져있다. 사무실 벽에 이러한 아이디어들과 모형들을 걸어놓고 이것들이 '프로젝트 디자인 가이드' 나 '디자인 방향' 이 된다. 그리고 나서 프로젝트가 끝나면 이 모든 것들을 수집해 보물처럼 상자안에 넣어둔다. (최종 수집을 위해 무엇이든지 쓰거나 그릴 때마다 프로젝트 이름과 날짜를 적어둔다.)

Our box-sketchbooks

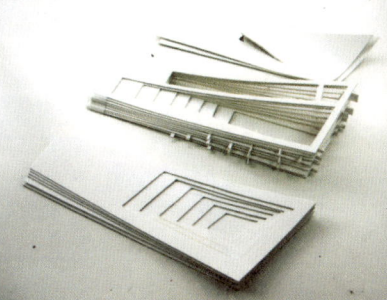

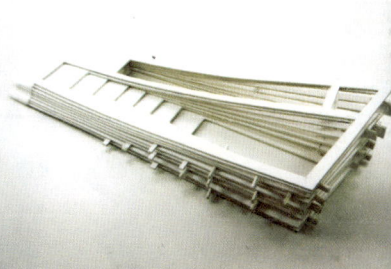

are the real project.

***Valer church**, Norway*

Jongyeon Bahk [Grid-A]

architour.pe.kr / grid-a.net

Partner of Grid-A Architectural Design, Joung Yeon Bahk also writes and manages a website about his drawings and information on architecture. The name 'Grid-A' is a combination of Grid, the basis of architecture, and the letter 'A'. Being the first letter of the alphabet, 'A' stands for 'the beginning', but can also be interpreted as the 'A' from 'Architecture'. When one reads it as one word, it sounds like 'to draw' ('gree-da') in Korean. This resembles the act of sketching and drawing architecture through illustrations, the works of this office.

What is a notebook to you?

It's a thought process. I record every moment of my thoughts in my notebook. During the school years, it was a book full of lecture notes. These days, it sometimes become presentation material to clients and team members when discussing about the projects throughout the design process. At times, it became my drawing portfolio, full of sketches and my thoughts and reminiscence on the memorable architectures that I visited. During travels, it also became my diary with admission tickets and receipts taped onto the pages.

당신에게 수첩이란 무엇인가?

생각의 흐름이라고 생각한다. 나의 수첩은 각 순간순간의 생각들을 담아내는데, 학부 시절에는 수업을 들으며 필기하는 노트가 되기도 했고, 건축설계를 진행하며 팀원들, 건축주와 협의 과정에서 설명을 위해 무엇인가가 그려지는 프레젠테이션 자료가 되기도 했다. 건축답사 중에 인상적인 건축물들을 스케치하며, 이 공간이 왜 아름다운지를 느끼기도 하며 드로잉 포트폴리오가 되기도 했다. 여행 중에는 입장권과 영수증들을 붙이며, 그 장소에서 느꼈던 것을 적어두는 다이어리가 되기도 했다.

우리는 건축공간을 형성하는 요소들을
처음에 눈으로 접하고 느낌을 갖는다.
"느낌은 대상에서 오는가? 아니면
내가 원래 가지고 있던 것을 대상이
일깨웠는가?"
　　　- 최부득 저. 건축에 있어서 정신적인것에 대하여

서른개의 바퀴살이 모여 한개의 바퀴통을 만들지만
수레를 움직이는 것은 가운데의 빈구멍

흙을 이겨 그릇을 만들지만
그릇을 쓸모있게 만드는 것은 그릇 속의 빈곳

방을 쓸모있게 하는 것은 그안의 텅빈 공간
그러므로 있음의 이로움은 없음의 쓰임에 있는것
　　　　　　　　　　- 노자 · 도덕경 -

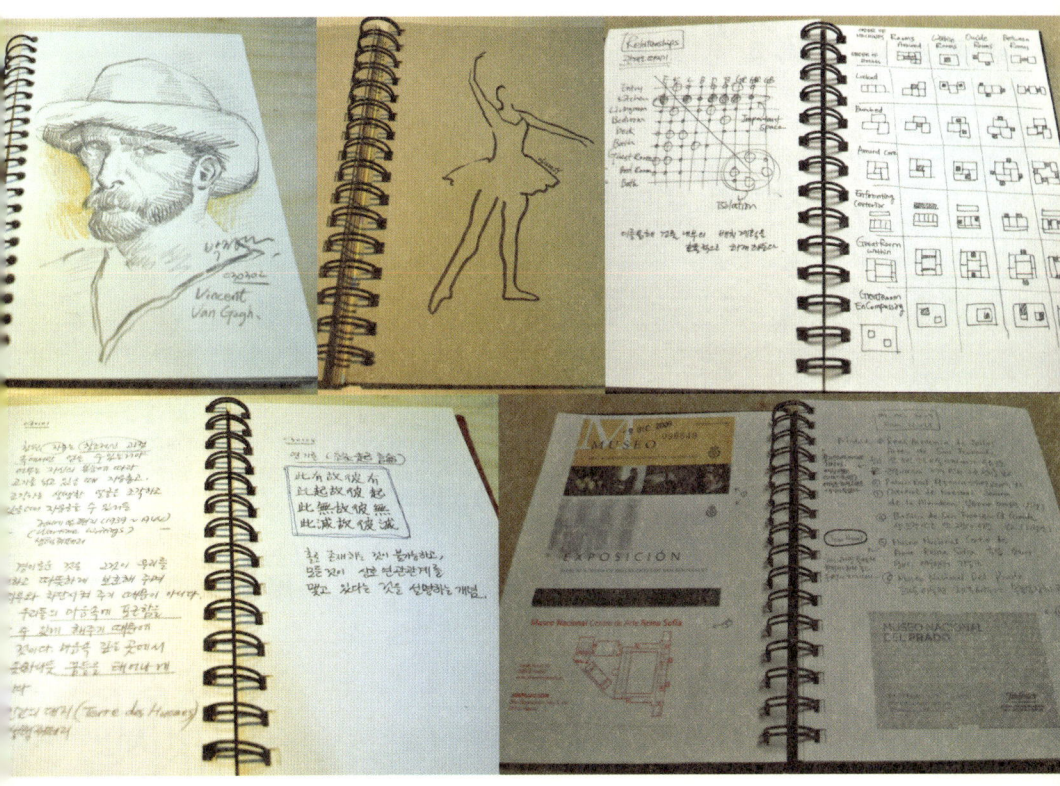

Any episodes or memories related to a notebook?

The most important aspect that I consider when purchasing a bag or going on a trip is whether my notebook can be easily carried and taken out for use. When I recently visited Niagara Falls, I was worried about my notebook getting wet, more than my camera or myself.

수첩에 관련된 에피소드가 있다면 들려달라.

가방을 구입할때나 여행을 다닐 때 가장 중요하게 생각하는 것이 내가 사용하는 수첩(스케치북)이 수납되는지, 쉽게 꺼낼 수 있는지 이다. 최근에 미국과 캐나다 국경의 나이아가라 폭포에 다녀왔는데, 폭포 속에서 카메라와 몸이 젖는 것보다 수첩(스케치북)이 젖을까봐 고심했던 기억이 있다.

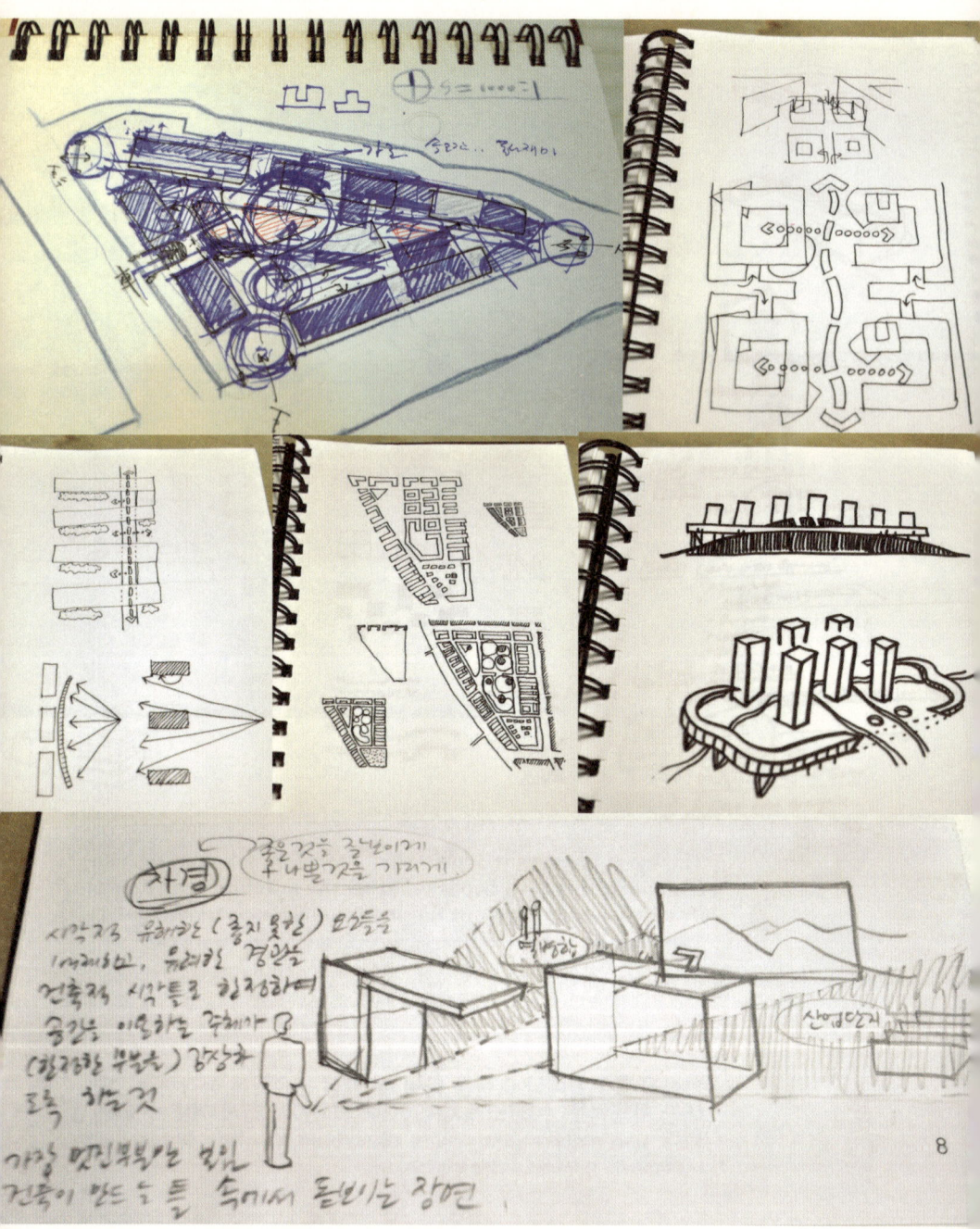

When and where do you use your notebook the most?

For the past couple of years, I've been sketching interesting spaces and architectures as I travel around a city or a country. I also use a sketchbook for every project that I work on and record all the process in it. So, I tend to use it all the time, since I have a sketchbook with me when I travel or work at the office.

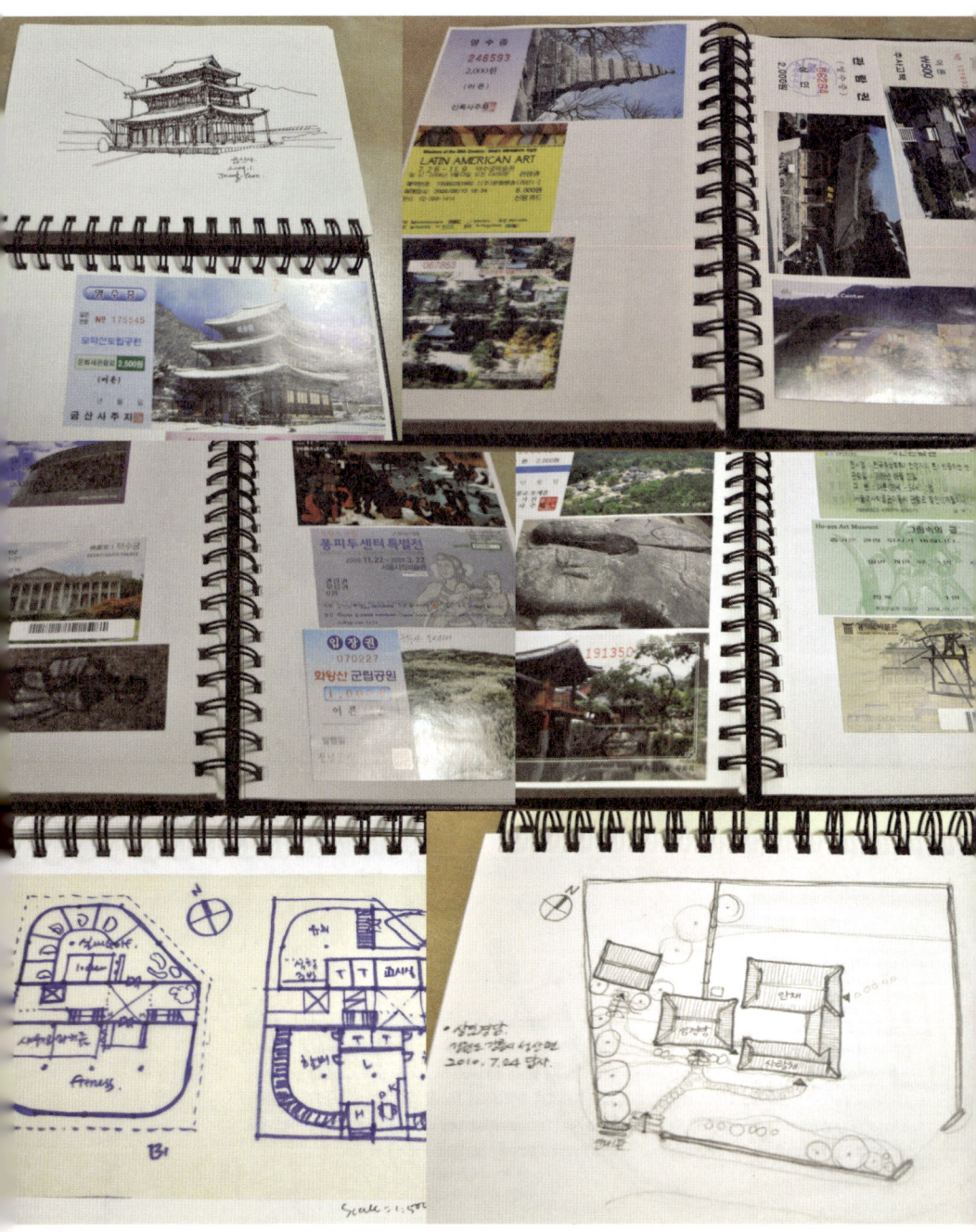

수첩을 가장 많이 사용하는 공간과 때는?

몇 년 전부터 한 도시, 혹은 한 나라를 여행하며 인상적인 공간과 건축물들을 스케치북에 담고 있다. 또한, 하나의 건축 프로젝트를 진행할 때마다 스케치북 한 권에 진행되는 내용들을 기록하고 있기에 여행/답사할 때, 혹은 사무실에서 일을 하며 꾸준히 사용하는 편이다.

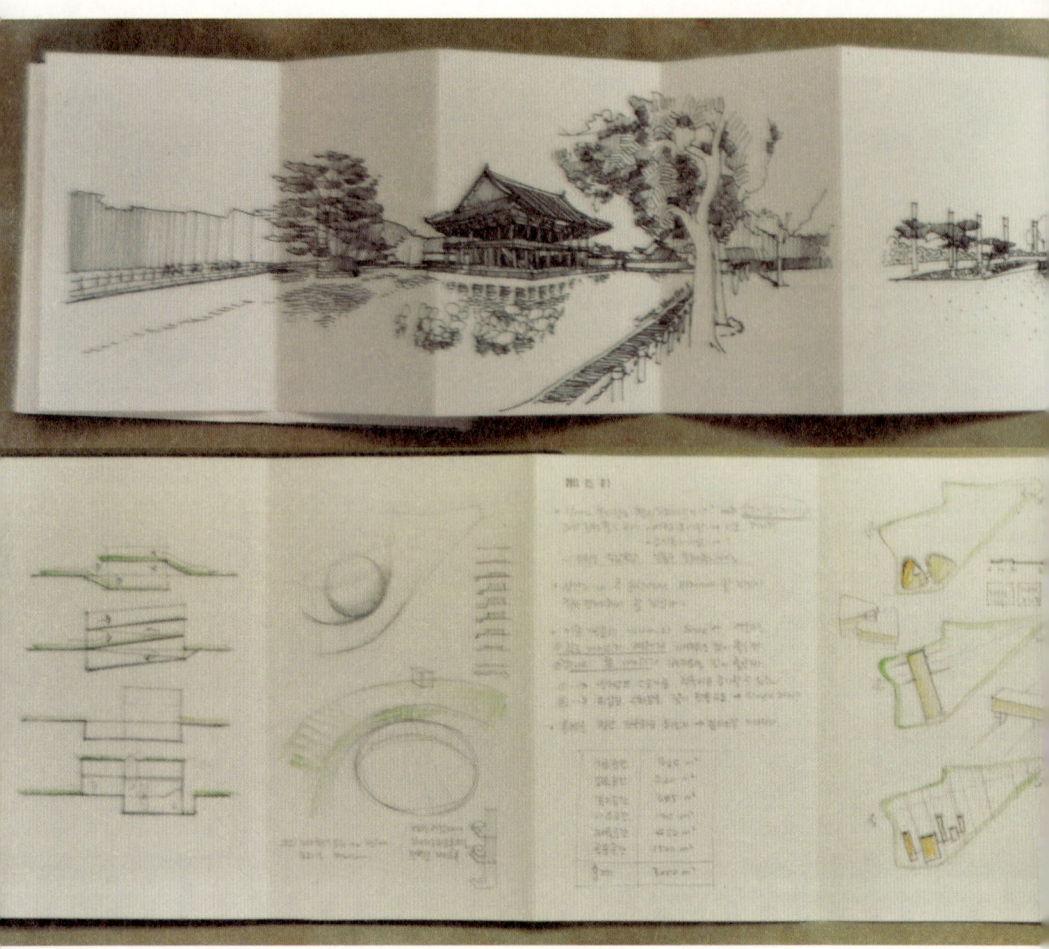

What influence does a notebook have in your projects and life as an architect?

Although one piece of paper is very thin, these pieces come together to make on book, and as numbers of these sketchbooks piled up, they became a large volume. Sketching and recording thoughts about architecture became a habit. As time passed by, I started to think to myself and define what beauty is and **what kind of space is a good space.** I think it is very important to know, as an architect, what is beautiful and how to design spaces. The habit of using sketchbooks helped me define myself as an architect who knows.

Are there anything else other than a notebook that you use to keep a record of your thoughts and ideas?

Although I use my smartphone sometimes, with all the various digital tools these days, I still like to write and draw with my hands. In between the pages of my sketchbooks, you can find odd pieces taped into them; they are papers from when I forgot to carry a sketchbook with me.

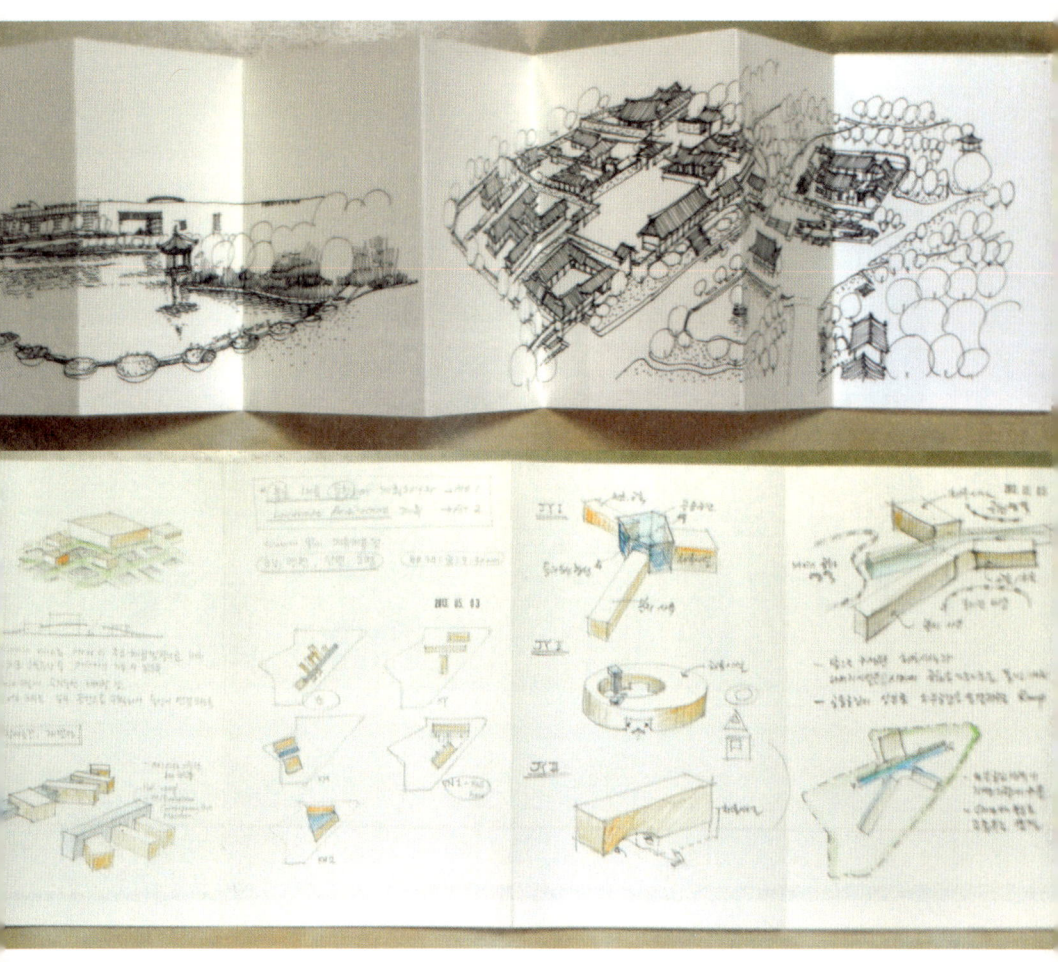

수첩은 당신의 작품과 삶에 어떤 영향을 끼치는가?

종이 한 장은 두께가 아주 얇지만, 수십 장이 모여 한 권의 수첩(스케치북)이 되고, 이것이 수십 권 모이니, 큰 부피가 되었다. 버릇처럼 건축물들에 대해 스케치하고 기록을 남기다보니 아름다움이란 무엇인가에 대해 스스로 정의해보게 되었고, **어떤 공간이 좋은 공간인지 생각할 수 있게 되었다.** 건축가로서 무엇이 아름답고, 어떠한 공간을 만들어야 하는지를 안다는 것이 매우 중요하다고 생각하는데, 이를 판단하는 기준을 갖게 된 것에 수첩(스케치북)을 사용하는 습관이 큰 기여를 했다고 생각한다.

수첩 외에 자신의 생각을 기록하는 방법과 도구는 무엇이 있는가?

많은 디지털 툴이 생겨나서 휴대폰을 이용하기도 하지만, 나는 손으로 적고 쓰는 것을 좋아한다. 수첩 중에는 수첩을 지참하지 못한 상태에서 다른 종이에 스케치하거나 메모해두었던 내용들을 오려 붙여둔 경우도 종종 찾아볼 수 있다.

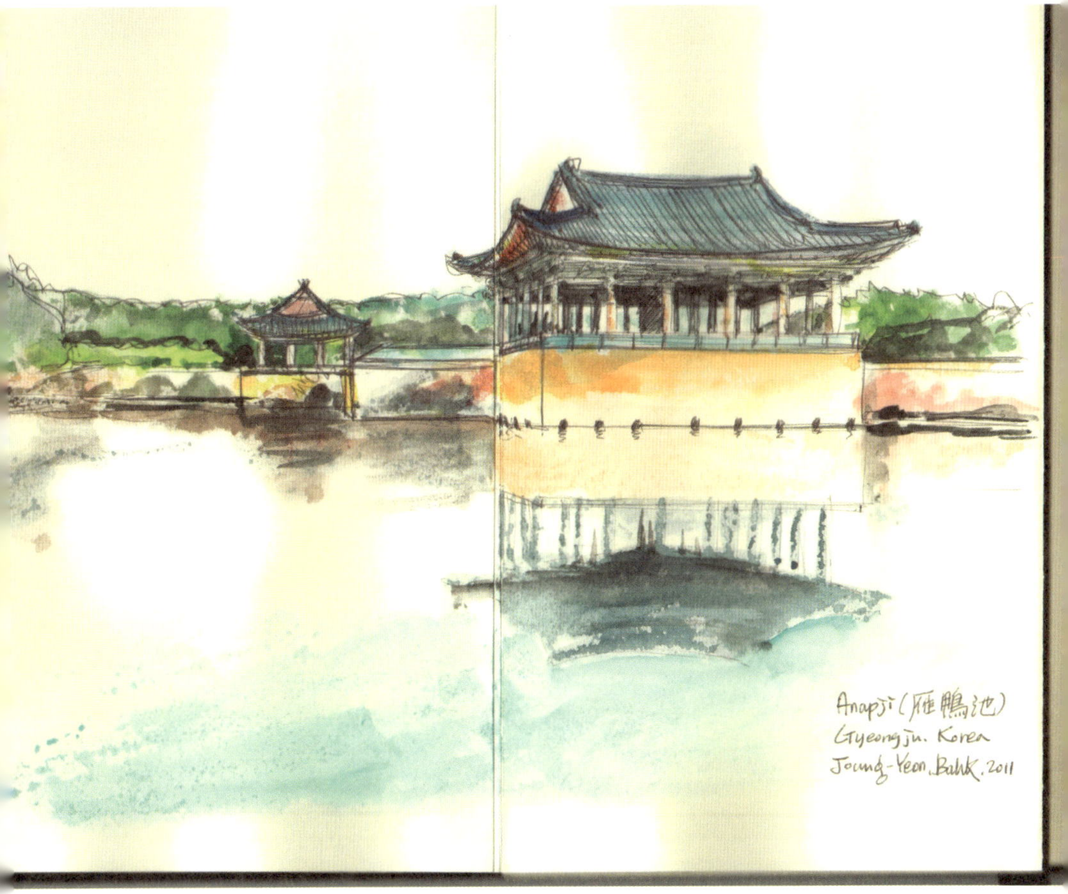
Anapji (雁鴨池)
Gyeongju. Korea
Joung-Yeon. Bahk. 2011

Russian Orthodox
Holy trinity Church
Joung Yeon. Bahk 2012

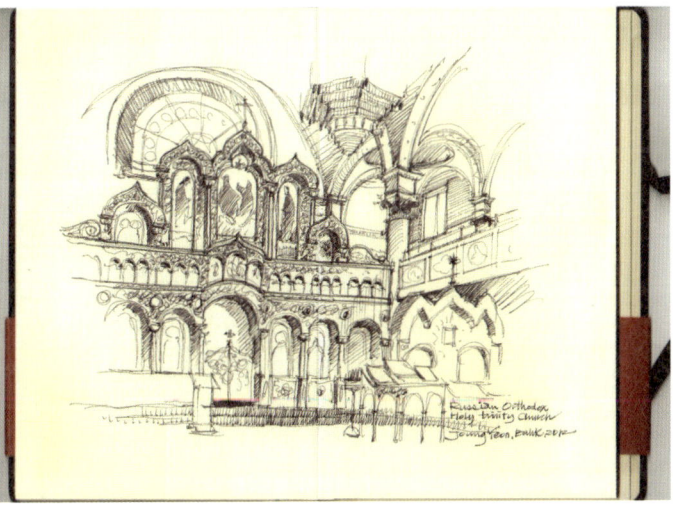

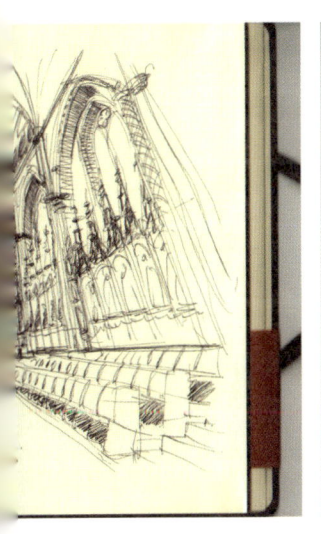

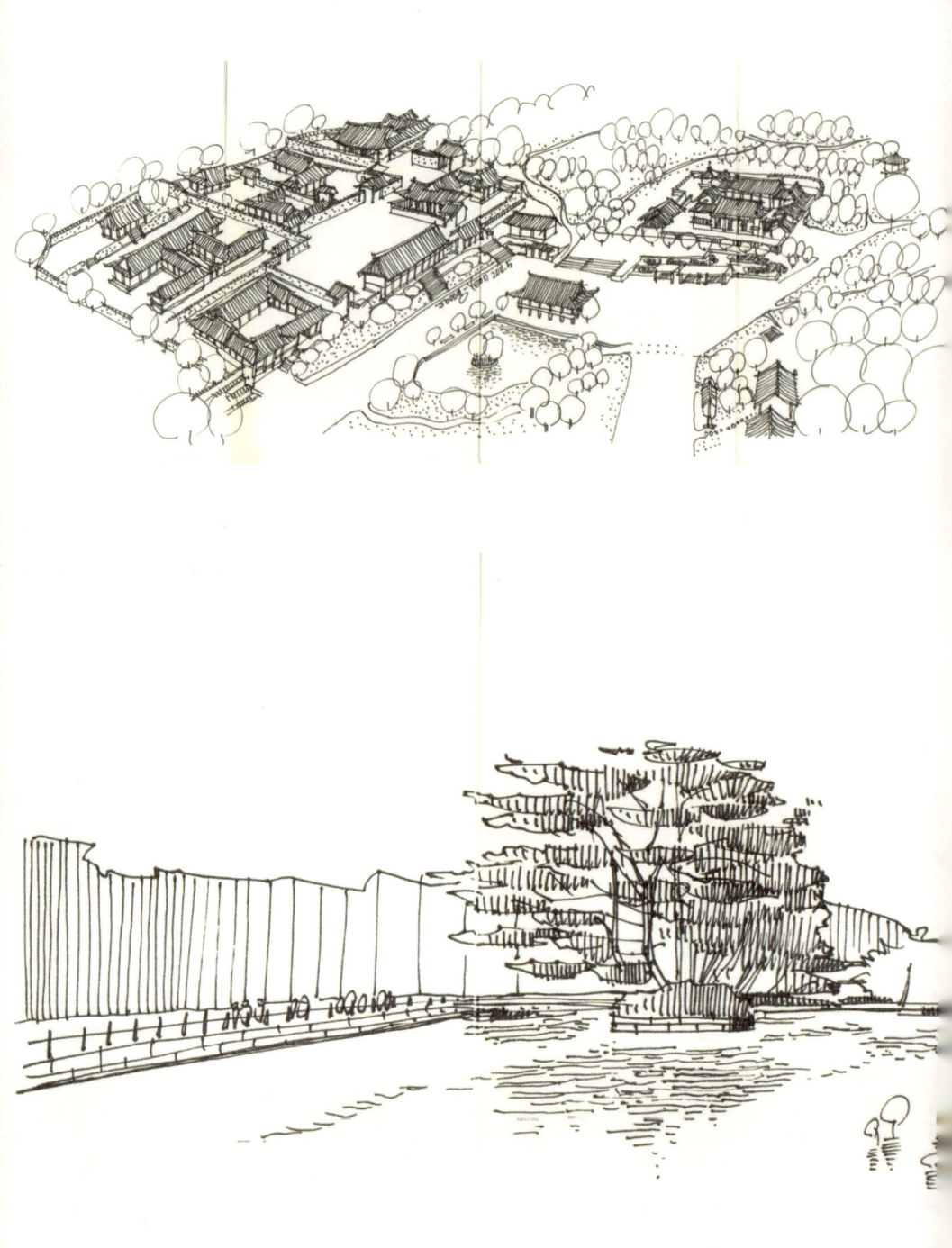

Joung-Yeon Baik 2012.

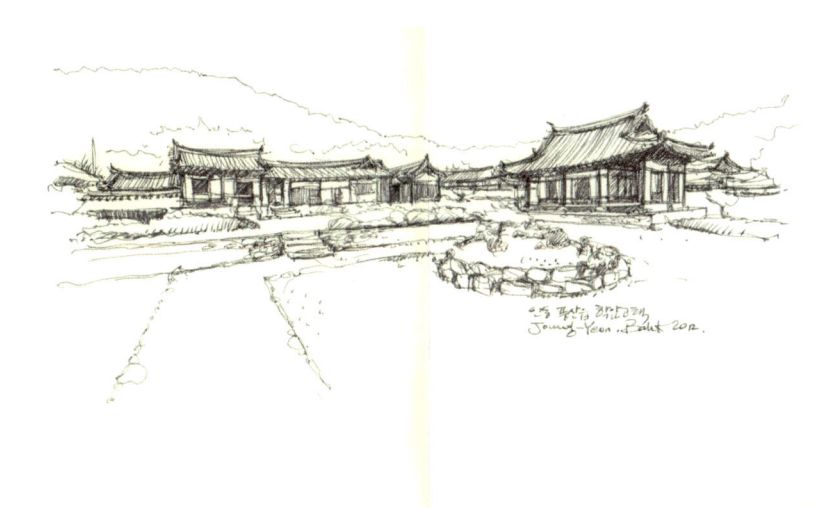

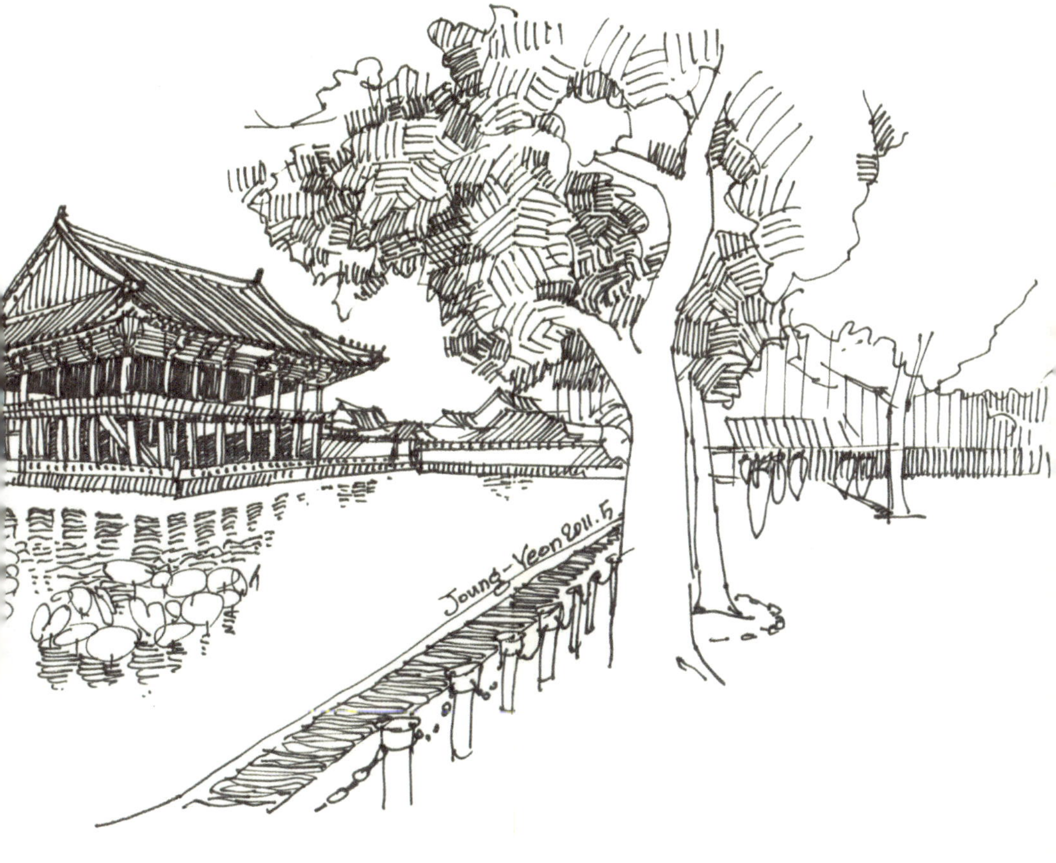
Joung-Yeon 2011.5

Cathedral Our Lady of Bur
부르고스 대성당. 유네스코 세계유산
으로 지정. 13세기부터 16세기까지
건축되었으며. 주 첨탑은 88m 높이,
예배당 내부 길이는 106m에 이른

Convento de San Esteban, Salama
대성당에서 걸음내려오면, 작은 광장을 지
브릿지로 연결되는 산에스테반 수도원이 있
16~17세기에 지어졌다고 하는데, 입면을
채우고있는 조각들이 인상적이었다.

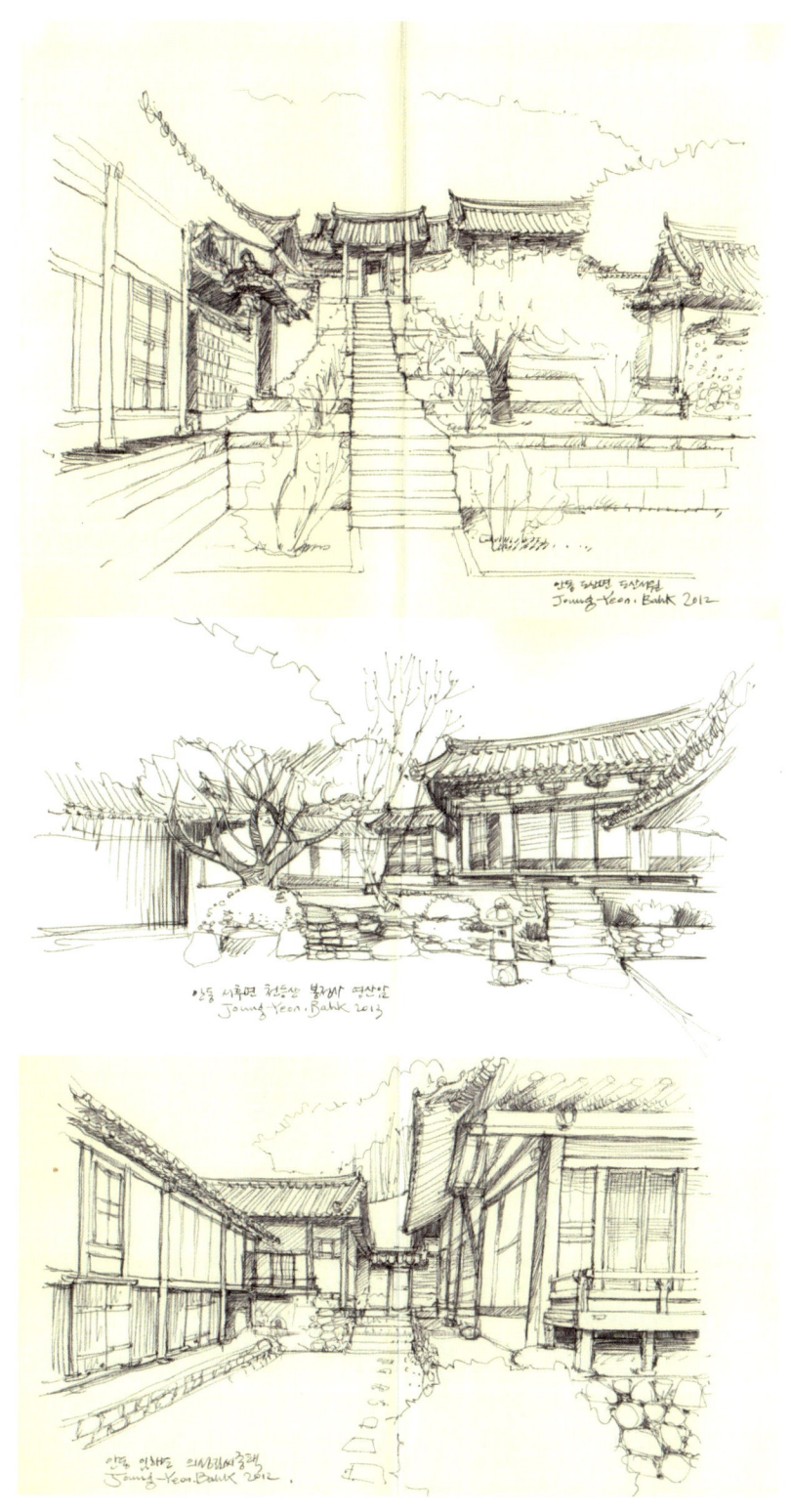

Omokdae 이성계 장군이 승전잔치를 벌인곳.
옛 전주 성읍을, 그리고 지금의 한옥마을을
한눈에 내려다볼 수 있는 곳. 전라북도기념물 16호.

江陵 船橋莊
강릉 선교장
20060630

김동수가옥 060609 JY

내진과 새진으로의 。05.2.2일차
마루그개 함과 벽에게 함.
Joung Yeon.

object-e architecture

object-e.net

Object-e architecture is a platform, created by Dimitris Gourdoukis in 2006, in order to explore new territories in architecture with the aid of computational tools and techniques. Object-e started as an exploration of new directions for design through computation; where computation is understood as the explicitly or implicitly coded digital processes that are aiming not in the computerization and optimization of existing practices, but in the invention of new ones. Through time, object-e moved beyond the borders of computation and engaged design at large, trying to graft computation with the social, political and ecological issues that architecture is facing today. Object-e is based on several collaborations with people coming from different backgrounds, with different design intentions and agendas. The outcome of this process, being in most cases collaborative, is therefore defying any concept of style; identity is formed through difference and constant transformation.

What is a notebook to you?

I use notebooks usually in order to record thoughts and ideas the time that they appear. **For me, most of the times, it works better with text than with drawings.** In that sense it is always good to have a notebook with me so whenever something comes to mind I can write it down. Usually I forget about what I wrote quite fast, but then after some time I go back to the notebooks, look through them and find again those ideas. In most cases they just stay there, in the notebook, but some of the thing might actually find their way into one of my projects or my texts.

당신에게 수첩이란 무엇인가?

나는 그때그때 생각나는 것들이나 아이디어들을 기록하기 위해 수첩을 사용한다. **그림이나 스케치보다는 대부분 글로 적는게 더 편하다.** 그래서 생각 날 때마다 적을 수 있게 수첩을 항상 가지고 다니는 것이 좋다. 내가 무엇을 적었는지 빨리 잊어버리는 편이기 때문에 스케치북을 다시 보면서 그 아이디어들을 떠올린다. 대부분 스케치북 안에만 남겨져 있지만, 어떤 것들은 나의 프로젝트나 글에 포함될 때도 있다.

For me, most of the times, it works better with text than with drawings.

When and where do you use your notebook the most?

Most of the times, I use notebooks when I don't have access to a computer. Long trips, boring lectures etc. Also when I need to escape from the computer and think and draw for a little while in a different way. **But in general, you can never know when you are going to need a notebook, so it is good to have always one with you.**

수첩을 가장 많이 사용하는 공간과 때는?

긴 여행이나 재미없는 강의를 들을 때 같이 주로 컴퓨터를 사용할 수 없을 때 수첩을 사용한다. 또는 컴퓨터에서 벗어나 다른 방법으로 생각하고 그릴 시간이 필요할 때 사용하기도 한다. **하지만 보통은 내가 언제 어디서 수첩이 필요할 지 모르기 때문에 항상 가지고 다니는 것이 좋다.**

What influence does a notebook have in your projects and life as an architect?

I am not really fond of that romanticized idea of the notebook, that it is the place where the architect creates his projects, how his ideas are conceived etc. It is the same concept of the paper napkin and the sketches you make on them. I don't believe that projects come out of sketches made in 1 minute on your notebook or any other piece of paper that you have available. My projects are conceived and developed through drawings, models, 3d models, scripts and simulations that most of the time happen on the computer and are the result of intense work, and a process that is developed over many iterations. The sketchbook is a tool that complements all the above. It is useful in order to record ideas in relation to the projects and processes that you are working on, but it is never the main instrument for them. It might be that their use is more psychological than really functional.

수첩은 당신의 작품과 삶에 어떤 영향을 끼치는가?

나는 수첩이 건축가가 자신의 프로젝트를 처음 탄생시킨 공간인 것 처럼 그렇게 낭만적인 것이라고 생각하지 않는다. 냅킨 한 장에 그린 스케치들과도 같은 컨셉이다. 수첩에든 어떠한 종이든지 간에, 1분만에 그린 스케치에서 프로젝트가 시작된다고 생각하지 않는다. 나의 프로젝트들은 도면, 모형, 3D 모델, 스크립트, 그리고 시뮬레이션을 통해 만들어지고 발전된다. 이것들은 대부분 컴퓨터를 사용하여 혹한 작업과 여러 단계를 거친 과정을 통해 만들어진다. 수첩은 이 모든 것을 보완해주는 도구다. 프로젝트와 그 과정에 관련되어 아이디어들을 기록하는 것에는 유용하지만 절대 주요 도구가 될 수는 없다. 실용적인 것 보다는 심리적인 것이 아닐까 싶다.

Are there anything else other than a notebook that you use to keep a record of your thoughts and ideas?

As I already said, I usually use digital computers to keep track of my ideas and thoughts, especially when they have to be expressed through drawings and images. When it comes to texts however, I use my notebook maybe to the same extend with my computer.

수첩 외에 자신의 생각을 기록하는 방법과 도구는 무엇이 있는가?

이미 얘기했듯이, 도면이나 이미지 들로 표현되어야 하는 것들은 주로 컴퓨터를 사용해 아이디어와 생각 들을 정리해둔다. 하지만 글을 써야 할 때에는 컴퓨터와 거의 동일하게 수첩을 사용한다.

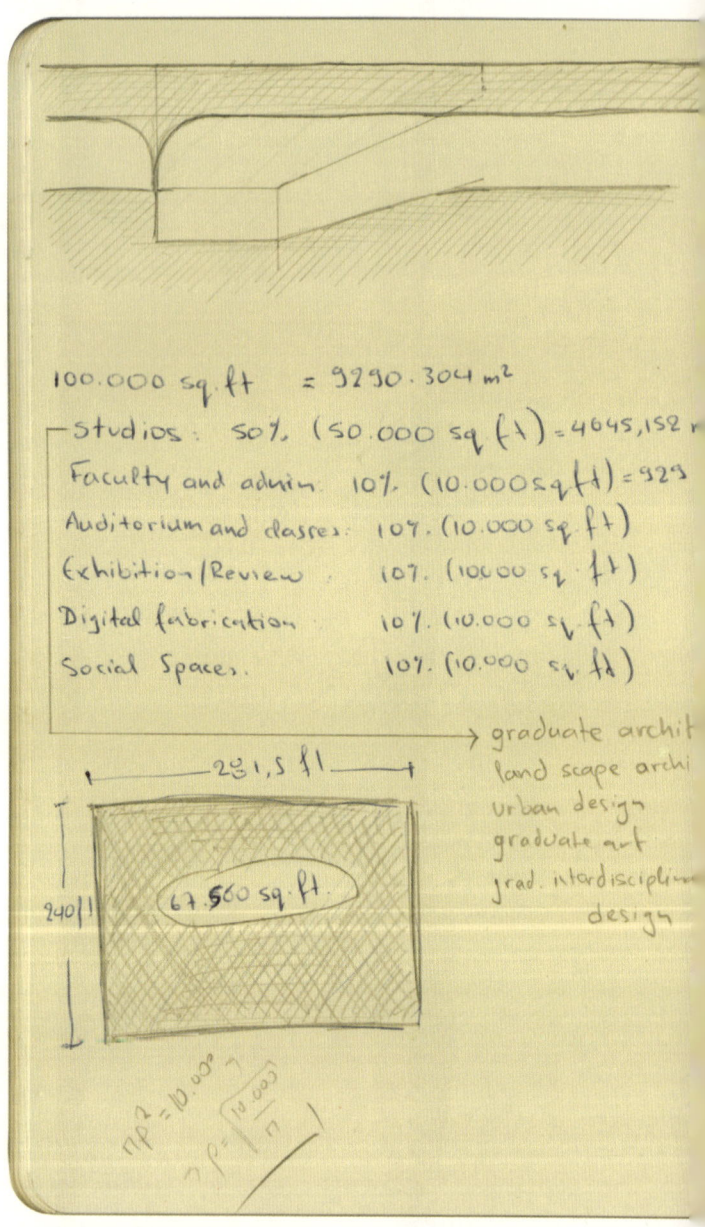

Hybrid

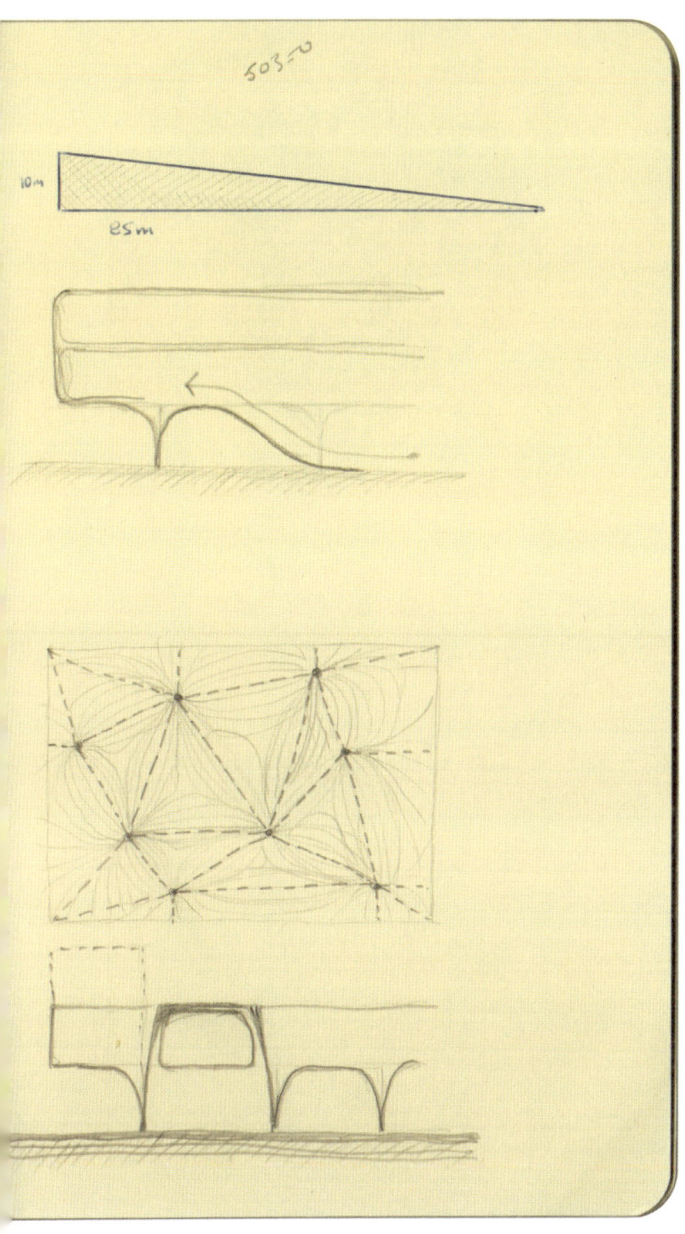

STUDIO
STUDIO
CLASSROOMS

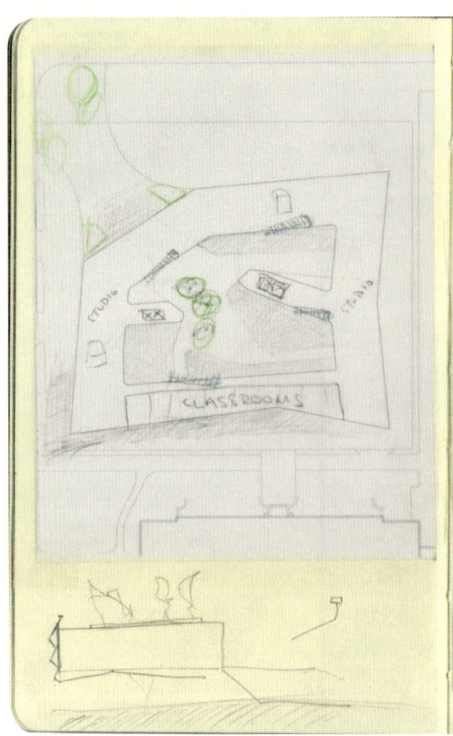

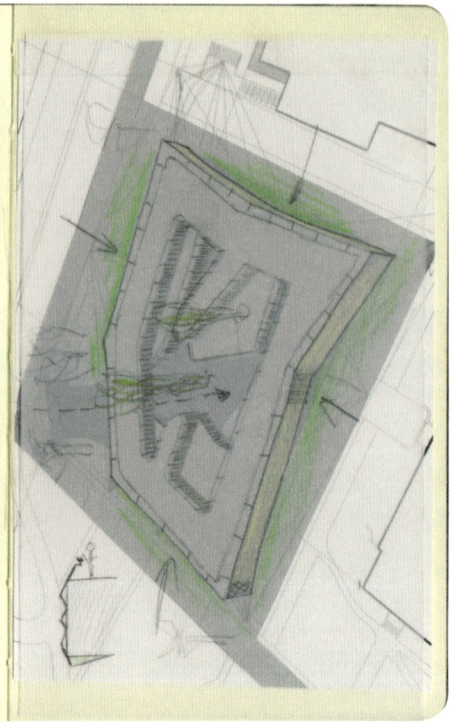

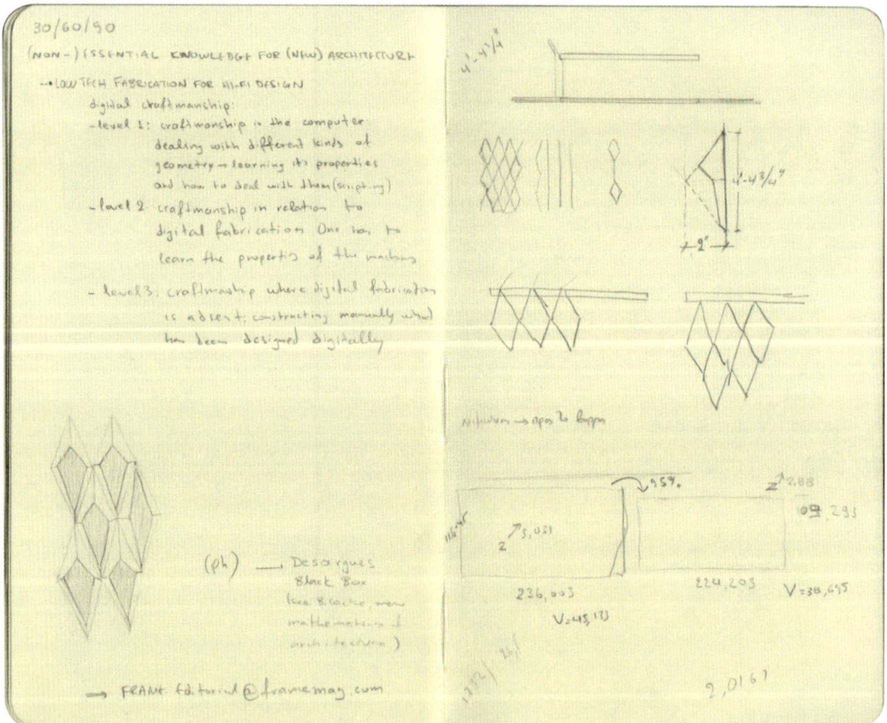

30/60/90
(NON-)ESSENTIAL KNOWLEDGE FOR (NEW) ARCHITECTURE
--• LOW TECH FABRICATION FOR HI-FI DESIGN
digital craftmanship:
- level 1: craftmanship in the computer
dealing with different kinds of
geometry → learning it's properties
and how to deal with them (scripting)
- level 2 craftmanship in relation to
digital fabrication One has to
learn the propertis of the machines
- level 3: craftmaship where digital fabrication
is absent; constructing manually what
has been designed digitally
(ph) → Desargues
Black Box
(see B. Cache, new
mathematics of
architecture)
→ FRAME editorial@framemag.com
4'-4 3/4"
2'
V=45,13
V=38,695
236,003
224,203
2,016

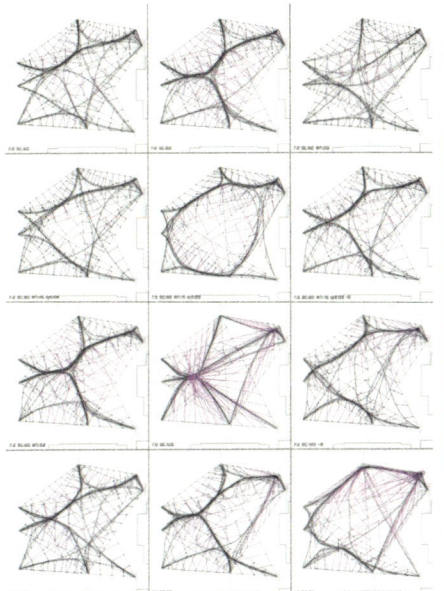

A superimposition of all the vectors displaying the displacement of the original routes

Different results of the simulation of the routes.

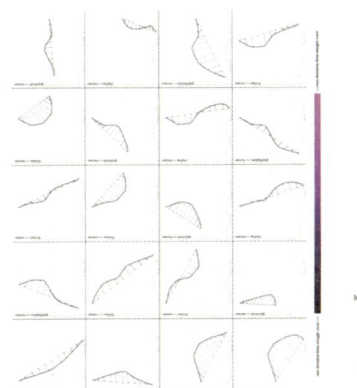

All the different paths accommodated from the solution and the deviation from the original paths.

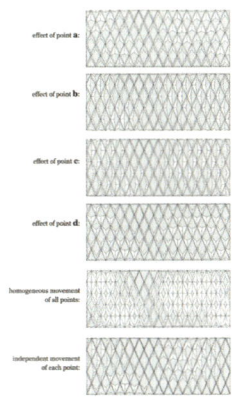

Results of the different components in elevation.

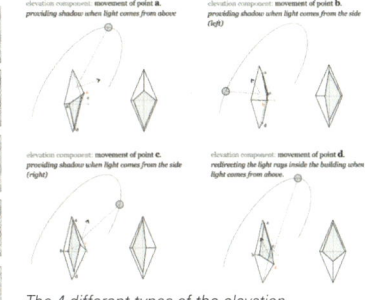

The 4 different types of the elevation components and how they deal differently with light rays.

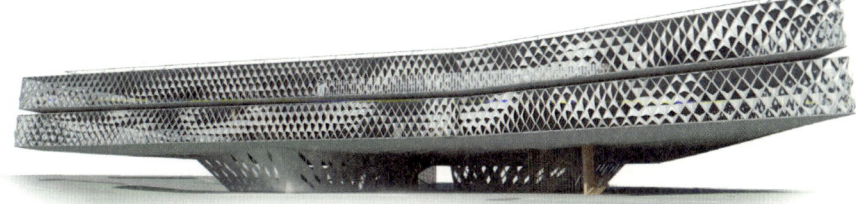

Distorted perspective view.

***L-SYS_HRRU.**
Design: Dimitris Gourdoukis

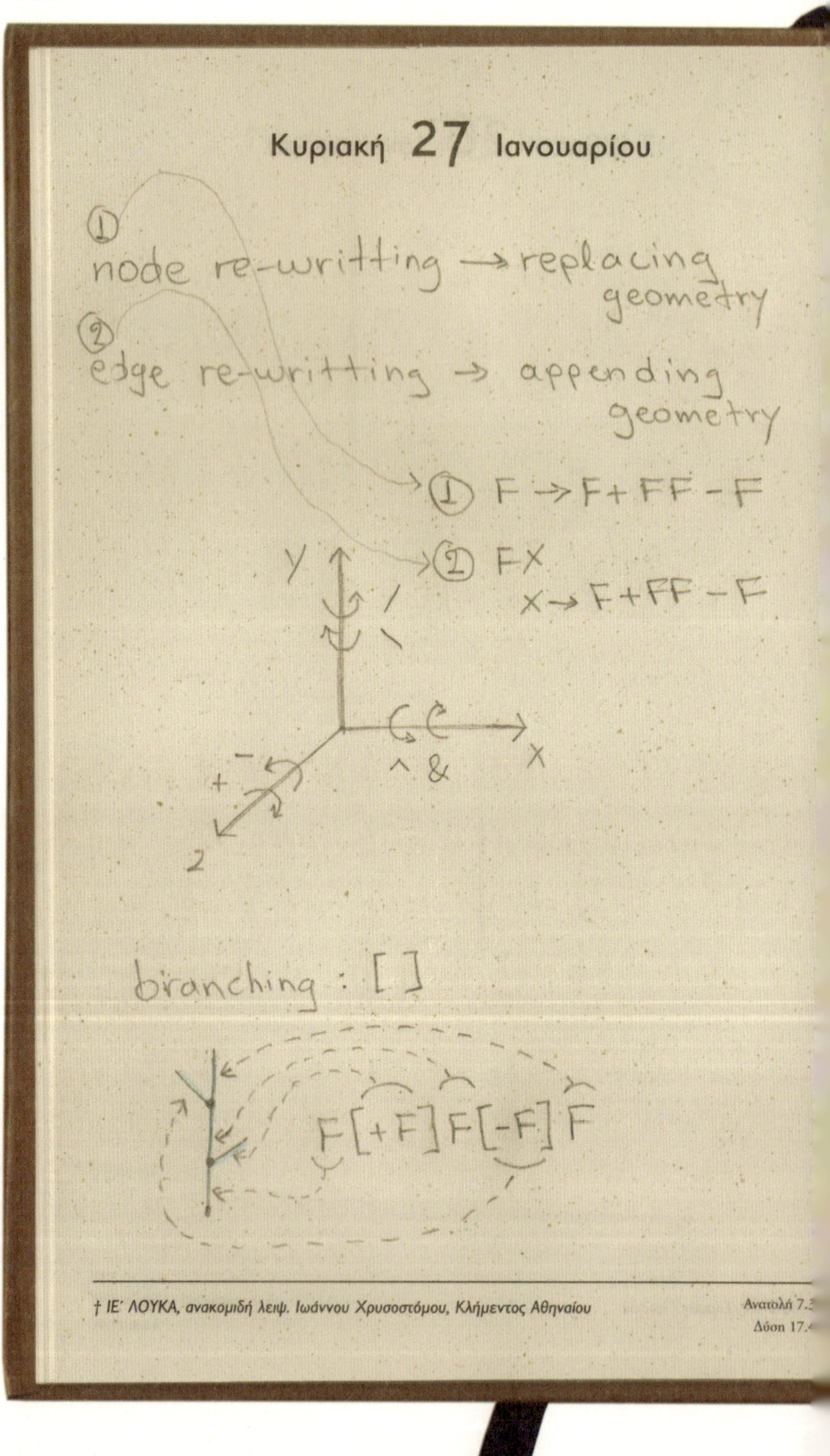
Κυριακή 27 Ιανουαρίου
① node re-writting → replacing geometry
② edge re-writting → appending geometry
① F → F+FF-F
② FX
X → F+FF-F
y
/
\
x
^ &
-
+
z
branching : [ ]
F[+F]F[-F]F
† ΙΕ' ΛΟΥΚΑ, ανακομιδή λειψ. Ιωάννου Χρυσοστόμου, Κλήμεντος Αθηναίου
Ανατολή 7.
Δύση 17.

Δευτέρα 28 Ιανουαρίου

( ) → overide values

to increase progressively:

$$X(h) = \dots\dots X(h+1)$$

$X : \underbrace{(t<4)}_{\text{condition}} = \dots\dots$

$X = \dots\dots\dots \underbrace{X : 0,5}_{\text{probability}} \quad (1:00)$

~(20) rotates randomly between 0-20

to stop this in favor of this

⬇

$A(h) : h<5 = FA(h+1)$
$A(h) : h=5 = A$

Εφραίμ Σύρου

Ανατολή 7.33
Δύση 17.43

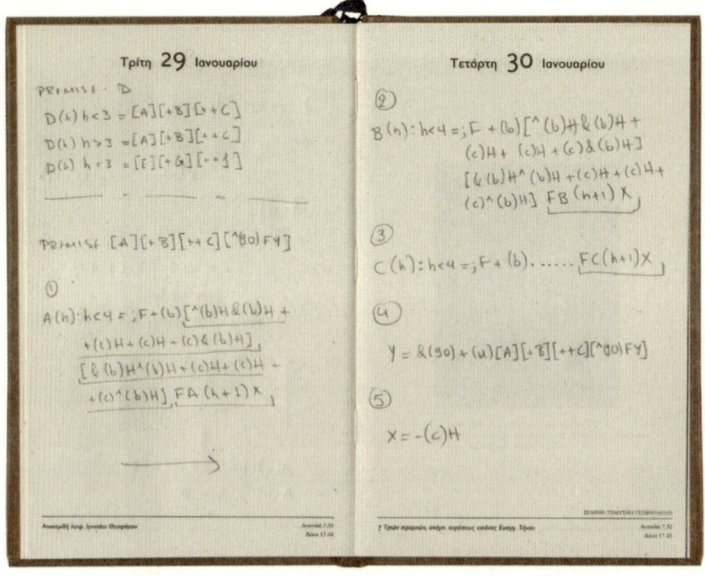

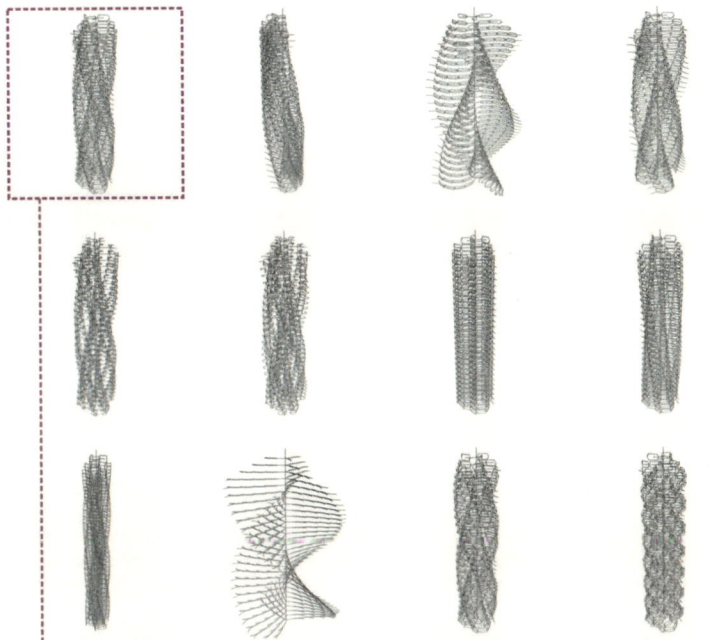

L-systems examples and various formations.

Premise: [A][+B][++C][^(90)FY]

rule01: A(h):h<4=;F+(b)[^(b)H&(b)H+(c)H+(c)H+(c)&(b)H][&(b)H^(b)H+(c)H+(c)H+(c)^(b)H]FA(h+1)X
rule02: B(h):h<4=;F+(b)[^(b)H&(b)H+(c)H+(c)H+(c)&(b)H][&(b)H^(b)H+(c)H+(c)H+(c)^(b)H]FB(h+1)X
rule03: C(h):h<4=;F+(b)[^(b)H&(b)H+(c)H+(c)H+(c)&(b)H][&(b)H^(b)H+(c)H+(c)H+(c)^(b)H]FC(h+1)X
rule04: Y=&(90)+(u)[A][+B] [++C][^(90)FY]
rule05: X=-(c)H

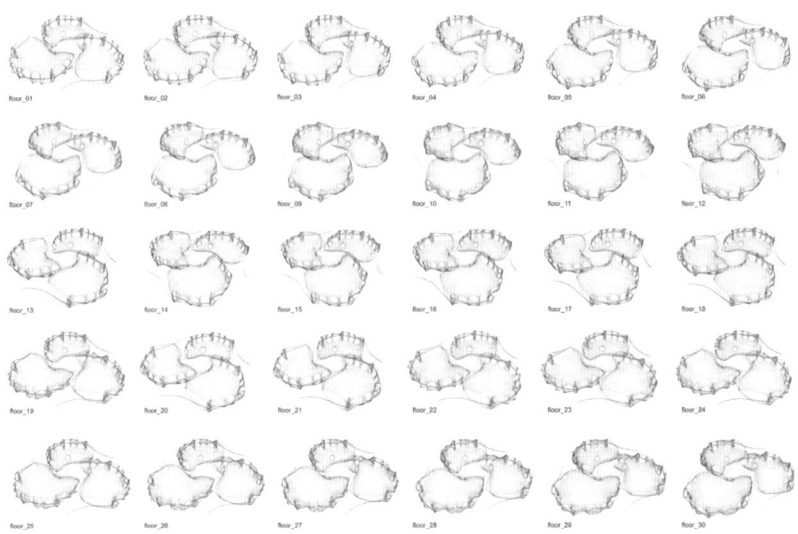

The geometry of all floor units.

Bird's Eye view of the tower.

Perspective view of the tower.

IaN+

www.ianplus.it

IaN+ was set up in 1997 and materializes around the core of its three members with diverse professional formation and experience: Carmelo Baglivo (30.11.'64), Luca Galofaro (19.03.'65), for design project and theory, Stefania Manna (06.07.'69) for engineering. IaN+ multy-disciplinary agency aims at being a place where theory and practice of architecture overlap and meet.

They have been participating to many national and international architectural competitions, gaining prizes and mentions. In 2006, with the "Tor Vergata" University scientific research building, they obtained the Italian Architecture Gold Medal for the first realized work at the Triennale in Milan, they participated at the London Biennale of Architecture and to the exhibition Talking Cities in Germany. In 2005 they were invited at the International competition for the new headquarter of the pharmaceutical company Angelini in Rome, winning the second prize, they got the third prize at the invited competition for the requalification of the Roman Theater in Spoleto and they participated to the exhibition "Avenirs de Villes/Future for Cities" in Nancy, France. In 2004 they won the international competition for the Tittot Glass Museum in Taipei, Taiwan ROC; they obtained the second prize for the international competition "New Ideas for the Daugava River embankment" in Riga-Latvia; the third prize for the international competition Tallinn Modules in Estonia.

Luca Galofaro's notebook

THE FORM OF INTERPRETATION-
note on conceptual form of representation.

Luca Galofaro

I own many notebooks, on each of them I recollect different things, I use This collection of images, photographs, collages and sketches to work, to develop ideas, to record data, to sketch architectural project, to visualize everyday reality. Recently I started to recollect those notebooks in a blog www.the-imagelist.com.
The main idea of the blog is to reorganize my file's image, starting from pictures found, taken, manipulated, architectures, drawings, collages some kind of visual auto biography. Most of the images are part of everyday work, models[1] utilized to think about a project, about architecture and spaces. A translating game around reality. A channel made in order to restore an esthetical code open to interpretation. The reorganization of the file needs different tools, keeping an independence but in the same time holding a relation between them . As a matter of fact I do not choose images with a clear logic, easily when a picture involve me I print it out and put it on a blank paper, where during the course of time I write notes. **I try to detect the image as possibility.** Sometime the same picture can be handle, cut, sample or become the basis for something new. Other pictures are realized on specific themes. Even on those, with time, I apply different alteration and become part of the file itself. The sketches are the fast writing of some ideas.[2]

Schetches
Writings, first sign of a future project, are the only non methodical gesture.
They follows the planning, come back in different time, the same sketches can follow more than a project or an idea.

Collages
Collages extemporary thought are born from non existing projects, from landscape calling architecture, from art. They experiment the sequential, the repetition and the object transformation, which in a short time become another thing, another space.
Other thought which take form on a slow period of time, grow and only sometime become the project.

Archive
The archive is composed by picture collected during the course of time, reorganized and systematized. The picture are work models, telling concepts and thoughts, research themes.
The model is an intellectual structure setting targets for our creative activities, just like the design of model-buildings, model-cities, model-communities, and other model conditions supposedly are setting directions for subsequent actions.

---

1) The model is an intellectual structure setting targets for our creative activities, just like the design of model-buildings, models-cities, model-communities, and other model conditions-supposedly are setting directions for subsequent actions. Osvald Mathias Ungers Morphologie. City metaphors.

2) The most beautiful pages of M. Houellebecq, without a doubt one of my favorite authors, are those inserted in The map and the territory, in this story the protagonist, unable to surrendering to nature, prefers his interpretation and cultural elaboration made by his own works, photographs, which have as their subject the Michelin maps. To the reality replaces then its interpretation, or rather its manipulation, so it starts the poetics of the author that acts as the interpreter of the world. Houellebecq know show to speak about our time as a few other writers, because dwells in it and is intrinsically permeated inside. A master in the analysis of the western production system, in the crumbling of the bodies, in relations and memory; he speaks about the reality that surrounds him proceeding with detachment and scientific minutiae, with the precision of a ethologist, out of any narrative process of magical-allusive type.
For the author, is the human in self which has become a mere commercial and cultural product, and therefore, a product of the art and culture ruthless market. The only solution for the artist, the last agent of the crafts, to which the industrial production has inflicted a fatal blow, is the way of solitude, because «life sometimes offers an opportunity, but when you are too cowardly or too indecisive to take that chance, she resumes her cards; there is a time to do things and to enter into a possible happiness, this moment lasts a few days, sometimes a few weeks or even a few months but only occurs once, only one, and if you later want to return to your own steps is simply impossible, there is no longer a place for the enthusiasm, the conviction and the confidence, they remain a sweet resignation, a mutual and saddened piety, the useless feeling that something could be, that's imply you seemed unworthy of the gift that you had been received»2.

This book is very well aware of my idea of poetry: the set of choices made by an author among all the possible artistic solutions regarding the purpose of his work, the relations with the tradition and the contemporary, the art of expression. These choices take on a meaning at the time when they come into contact with the world and are assimilated before and interpreted after.
In asensealso the architects' poetics is the result of a mediation, of a selection of information, of a story. The works of Jed Martin, the protagonist of the novel, are described but never represented, the text replaces the image or better suggests the formation of the images, which take body in the mind of the reader (to illustrate this text I have inserted some, a my free interpretation, of course).
The pictures are trying to give poetic form as described by Houellebecq3. By an interpretative operation the artist take the distances to the world of reality and makes us to penetrate his world, his view defines an other-reality that re-builds through the signs the territory.
At the moment in which is represented by cartographer, the territory is enriched with meanings, that the nature, in reality, confounds. The view of the artist is indeed more true of the reality it self. «[...]while the satellite photo let appear only a mixture of greens more or less uniform, strewn with vague blue dots, the paper developed a fascinating entangle of provincials, of pictures queroutes, of points of view, of forests, of lakes. Above, with black capital, was the title of the exhibition -The map is more interesting than the territory4».
The poetics isn't born from a subject's aware operation but by the free interpretation of his work by its users. An author, an artist or an architect doesn't define their own poetics as an autonomous action, but as a result of a system of relationships that establish with others and with the reality that want to represent.
Each reading is born from a analysis, and from an interpretation of the historical time in which the work has form.

해석의 형태 – 표현에 대한 개념적인 형태

Luca Galofaro

나는 각각 다른 종류의 수첩을 여러 권 가지고 있다. 나는 아이디어를 발전시키고 정보를 기록하며, 건축 프로젝트를 스케치하고, 일상생활을 시각화하기 위해 이미지, 사진, 콜라주, 그리고 스케치들을 모은다. 요즘에는 내가 가지고 있던 수첩들을 블로그에 옮겨 담기 시작했다(www.the-imagelist.com).

블로그의 중점은 내가 가지고 있는 이미지 파일들을 다시 정리하기 위해서다. 내 블로그에는 내가 찾았던 것, 찍었던 것, 변형시켰던 것, 건축물 사진들, 그림들, 콜라주들 등이 있다. 대부분의 이미지는 매일 작업하는 것들의 일부분이다.[1] 그중에는 프로젝트와 건축 그리고 공간에 대해 생각하기 위해 만들었던 모형들도 있다. 현실을 바꾸는 게임. 다양하게 해석할 수 있는 미적인 것들을 회복시킬 수 있도록 만들어진 것. 파일들이 독립될 수 있도록 각각 다른 방법으로 정리해야 하지만, 동시에 서로 관계를 유지해야 한다. 솔직히 말해서 난 어떠한 논리를 가지고 이미지들을 고르지 않는다. 노트를 쓸 때에 필요한 이미지를 프린트 한 후 흰 종이에 붙인다. **나는 이미지를 가능성으로 보려고 한다.** 가끔 같은 이미지는 자를수도, 샘플이 될 수도, 또는 또 다른 새로운 것의 바탕이 될 수도 있다. 특정한 테마로 현실화되는 이미지들도 있다. 이러한 이미지들조차도 시간이 지나면서 다양한 변형을 시킨다. 스케치는 아이디어를 빠르게 적을 수 있는 방법이다.[2]

스케치

미래의 프로젝트에 대해 유일하게 체계적이지 않은 것은 바로 글이다. 계획을 따라가다가도 다른 시간에 돌아올 수도, 같은 스케치가 여러 프로젝트나 아이디어를 표현할 수도 있다.

콜라주

즉흥적으로 생각을 모으는 콜라주는 존재하지 않는 프로젝트로부터 만들어진다. 랜드스케이프가 될 수도, 건축이나 예술이 될 수도 있다. 콜라주는 짧은 시간 안에 다른 것이 될 수도, 다른 공간이 될 수도 있는 오브제가 순차적으로 변하고 반복되는 것을 실험한다. 오랜 시간 동안 형태를 만들어가는 것은 점차적으로 발전되어 가끔 프로젝트가 되기도 한다.

기록 보관

아카이브는 시간이 지나면서 모아 재정리하고 체계화된 이미지들로 이루어져 있다. 이미지들은 작업 모델로서 컨셉과 생각 그리고 연구했던 아이디어들을 알려준다.

모델은 우리의 창의적인 활동 목표들을 만들어주는 지능적인 구조다. 모델-빌딩, 모델-도시, 모델-지역, 그리고 다른 모델 조건들의 디자인은 차후의 행동들을 위해 방향을 알려주는 것과 같다.

Sketches

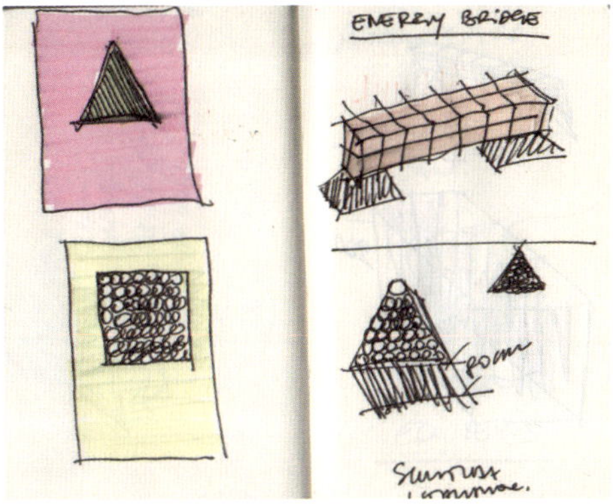
ENERGY BRIDGE

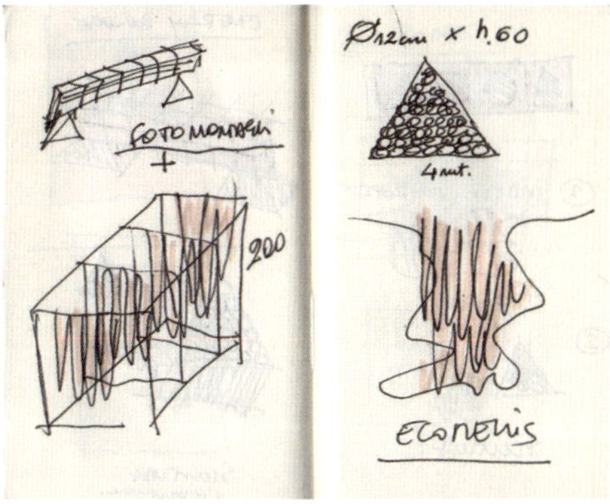
200
4mt.

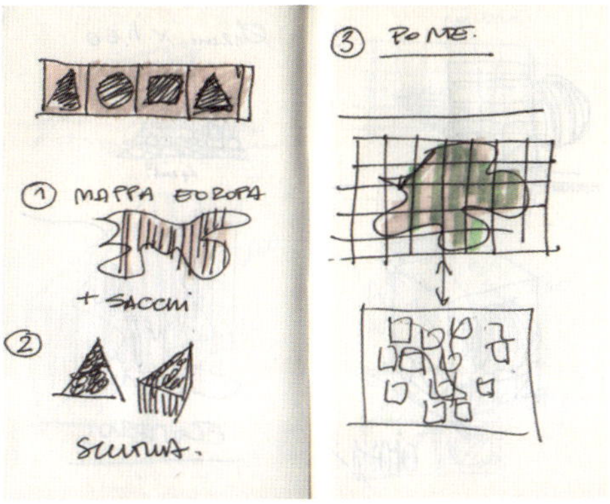
1 MAPPA EUROPA
+ SACCHI
2
3 PONTE

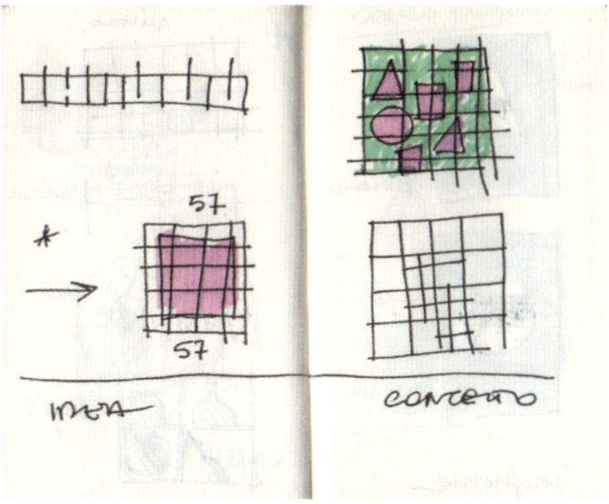
57
57
IDEA
CONCETTO

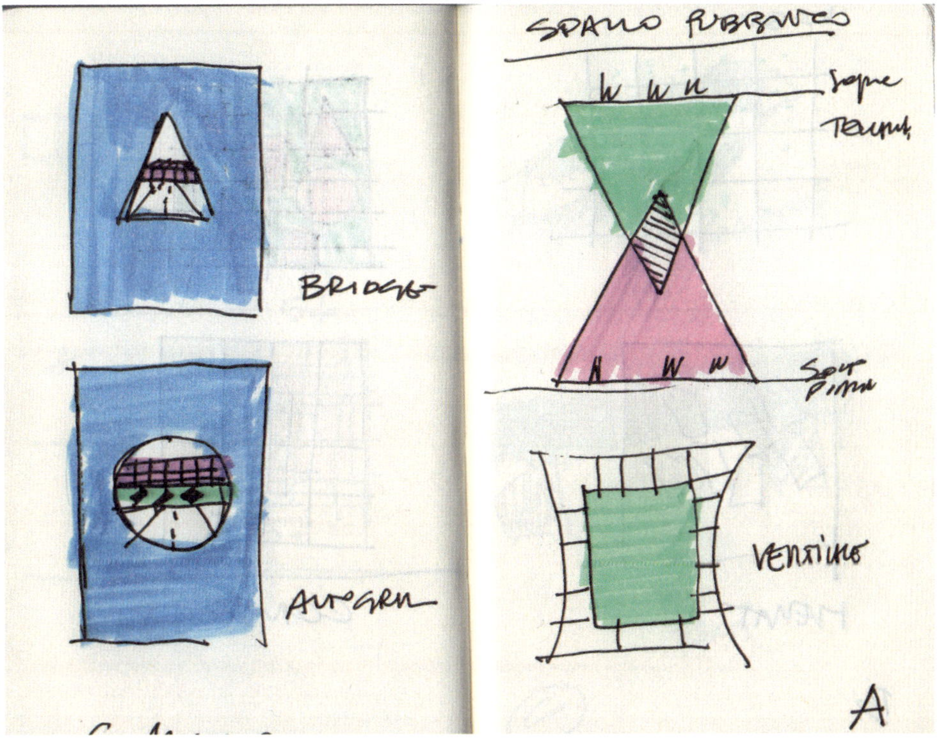
SPAZIO PUBBLICO
BRIDGE
A

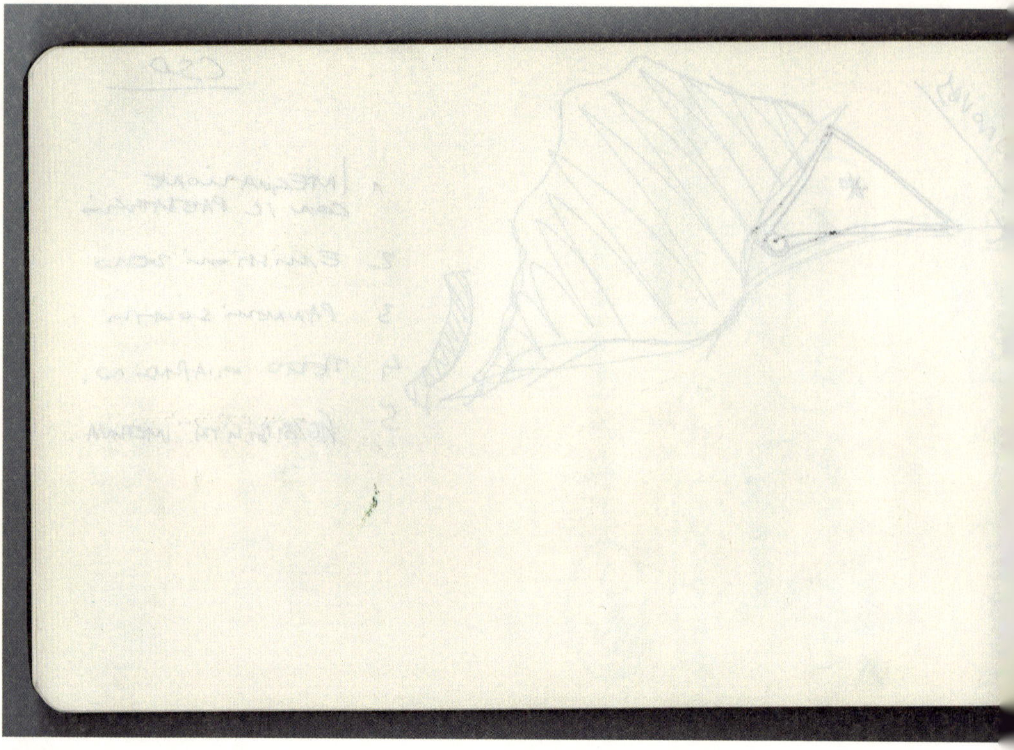

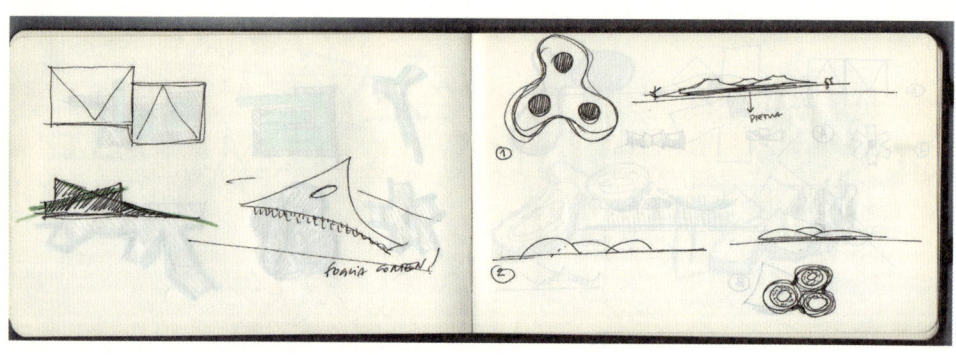
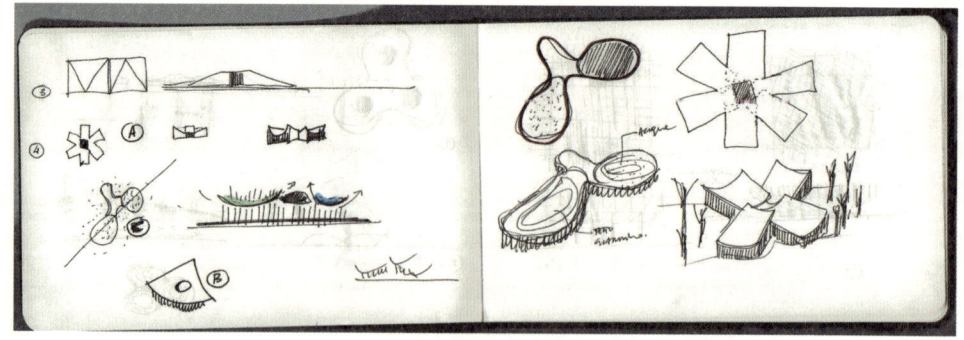

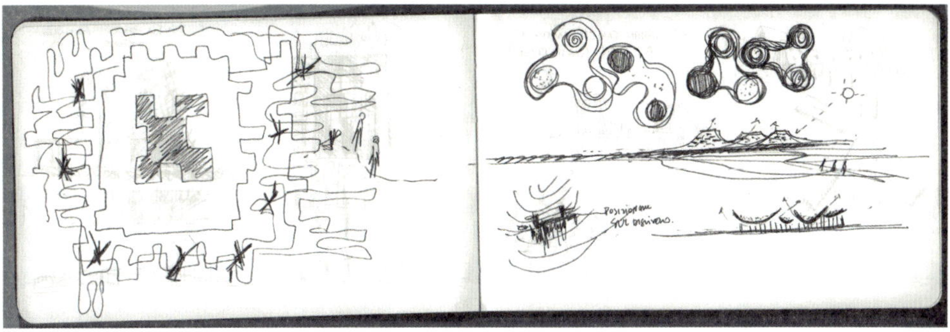

323

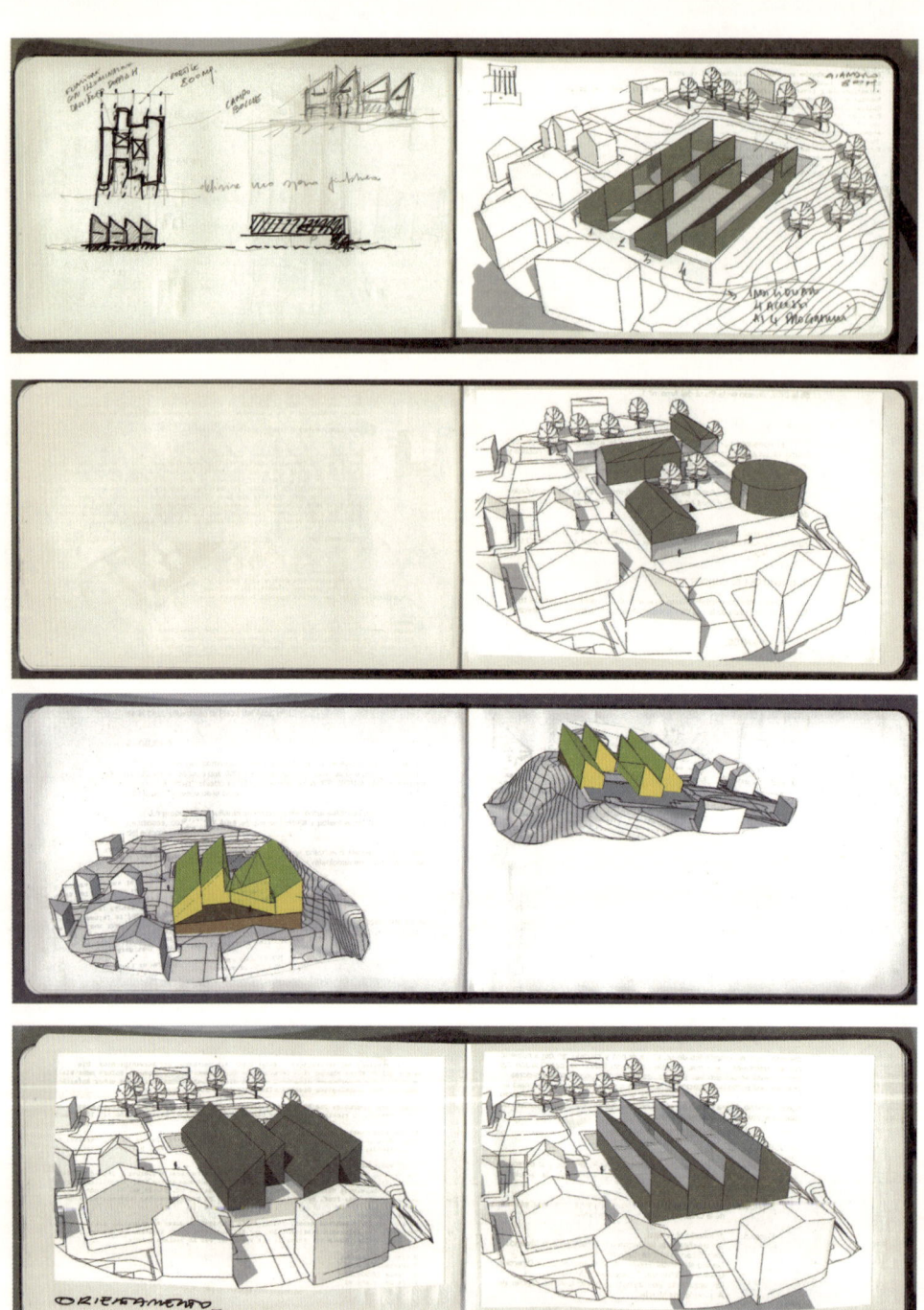

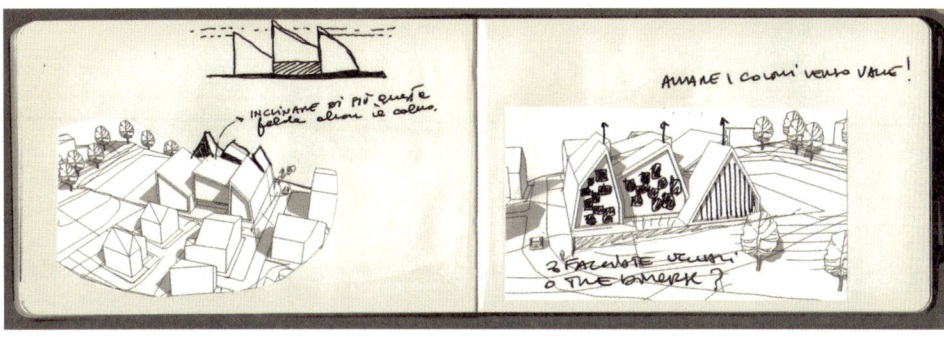
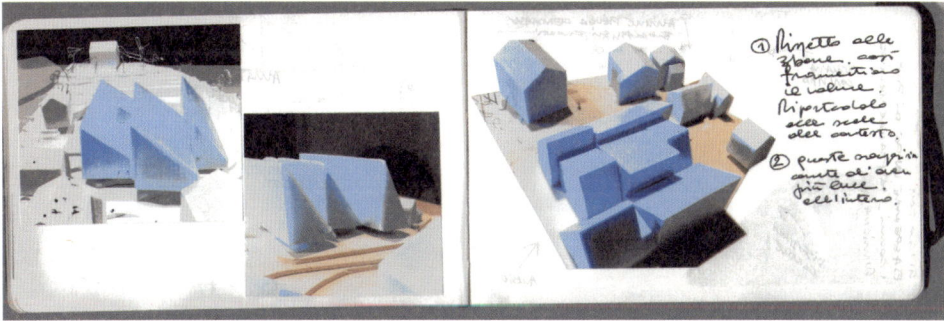

© **Sketches**, Luca Galofaro

Piramide Cestia Roma

Interpretazione - Omaggio a Piranesi

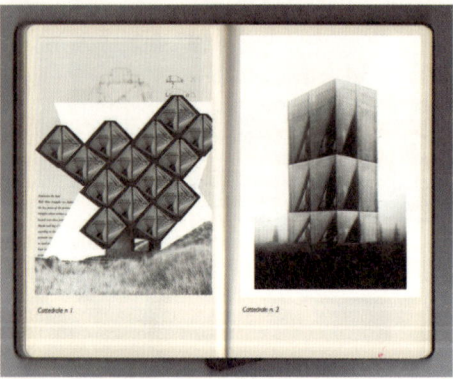

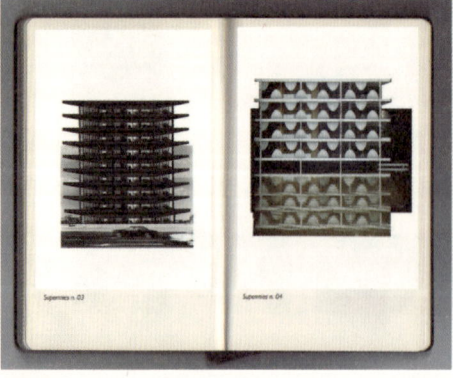

Variazioni Piramidali n.3 — Variazioni Piramidali n.4

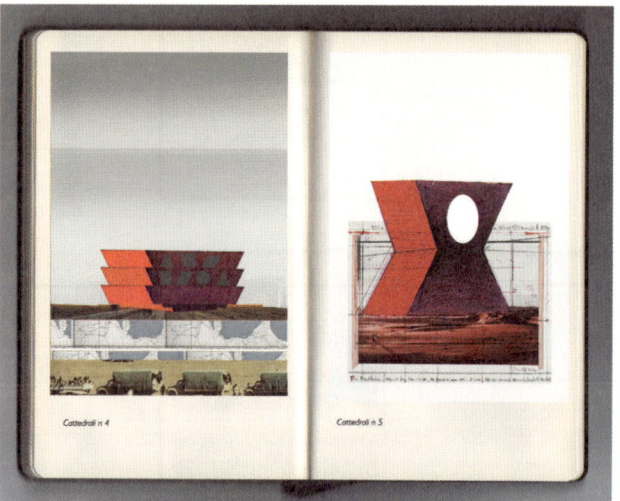

Cattedrali n.4 — Cattedrali n.5

Infrastruttura n.1 — Infrastruttura n.2

Archive

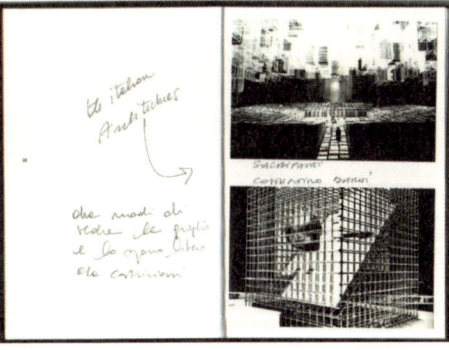
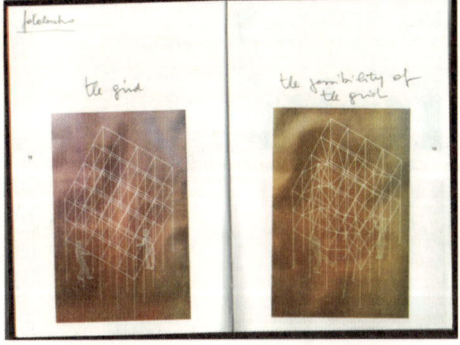

33
Francesco Arena

the grid and the form

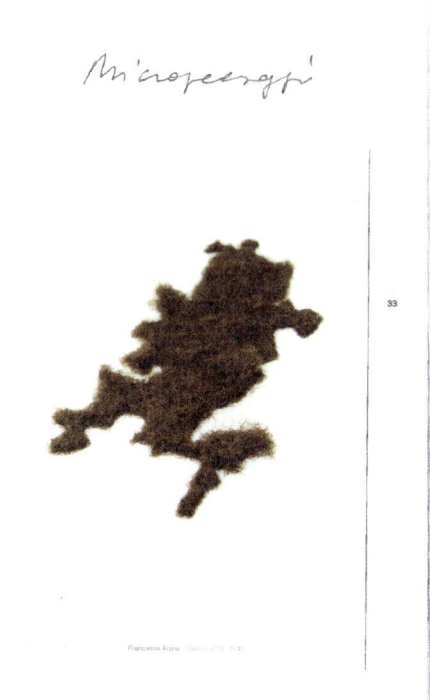

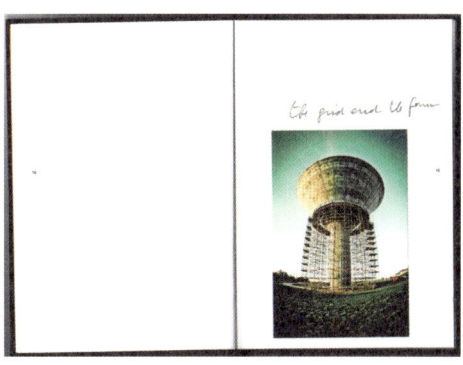

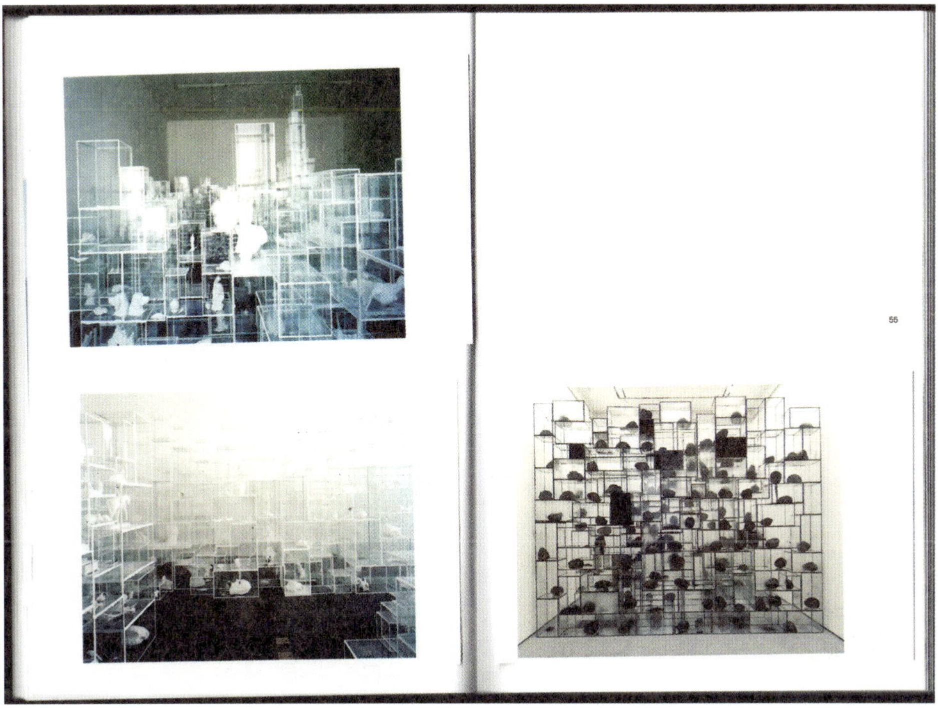
55

Carmelo Baglivo's notobook

THE FORM OF ARCHITECTURE: SUPER CENTRE POMPIDOU

Carmelo Baglivo

From the Otolith Group exhibition "Thoughtform": "Thought-Forms" a book published in 1901 written by Annie Wood Besant and Charles W. Leadbeater: "... ...to live we must know, and to know we must study; and here is a vast field open before us..." there is a difference between thinking and reasoning, because thinking implies understanding the meaning beyond what can be expressed in words ...
... Changing, adapting, becoming different, resembling, echoing themselves ... all the works must be able to turn into another form, even to express things in a different way, to be represented starting from different premises."

To work with what we have, with the city thought as a whole organism and not divided between center and periphery. On the contrary to work in the center, as a place to re-establish and re-urbanize.
The city at the center of the research, the city as a monument, the monument as an imposing presence but, a great absence, a great void as well. Monumental void.
Get away for a moment from the reality of practice to get into the reality of provocation.
Look for new development models and a new image of the city. Through images think of new models and new paradigms: but which images and models?

Otolith again wrote: "for us there is no memory without image and no image without memory. Image is the matter of memory".
This does not mean the way we regard the past but the way we think about the future, the way we imagine it, and then: What is an image? Can it be independent from the context?
Deleuze "the thought in thought " as a result of an image-shock, if ideas may be unclear images will be always clear.
Super Beaubourg represents a new logic of modernity where buildings are to be implemented, not iconic.
Buildings like structure, difficult to name.
Baudrillard defines the Beaubourg: "a monument to mass-culture that absorbs and grows, the mega cultural device resolves the relationship between inside and outside absorbing and reinterpreting the culture, the history and the city. It is the history that changes and tells the world. "
In the logic of a new modernity these buildings grow and incorporate everything, the city grows on itself, is recycled, it but does not devour new territories but it saturates and consumes itself.
Rem Koolhaas: the city cannibalizes itself!

Oswald Mathias Ungers declares, "that his buildings are neither romantic nor rationalist, neither traditional nor modern, but they try to connect strongly to the reality of the place, to the reality of those who live in that place and its history.

Schopenhauer: "Concept is like a dead tank where, what has been placed inside it is really all together, but where it is also impossible to take out more than what we placed; the idea instead develops representations that, compared to its same concept are totally new, it's like a living organism that produces what is not canned".

OMU says: it is impossible to think without a mental image.

Now architecture is totally self-referential, it self generates knowledge; architects do not need anymore the city as the place of representation and knowledge, but they need media. This has widened the field of action and representation approaching architecture to that Bauman's world liquid, but at the same time has dismissed architecture from being an autonomous discipline, regulated by laws of architecture and history; and that was where the social value of architecture resided, while now it resides in the consumption of the places, that is only in their use.

건축의 형태: 퐁피두 슈퍼 센터

Carmelo Baglivo

1901년도에 애니 우드 베전트와 찰스 리드비터가 1901년도에 출판했던 Otolith Group 전시 책 'Thoughtform': 'Thought-Forms'에 일부분이다. "… 살기 위해서는 알아야 하고, 알기 위해서는 공부해야 한다; 이 때 우리 앞에는 넓은 들판이 놓여있다… 생각하는 것과 추리하는 것의 차이는 생각 하기 위해서는 말로 표현될 수 없는 것을 이해하고 있다는 것이 전제된다.

변하고, 적응하며, 달라지고, 닮아가며, 반복하는 것… 모든 것은 다른 형태로 변할 수 있어야 한다. 다른 방법으로 표현할 수 있어야 하고 다른 전제로 시작되는 것을 나타낼 수 있어야 한다."

우리에게 주어진 것으로 작업할 때에는 도시를 중심과 주변부로 나눠서 생각할 것이 아니라 하나의 유기체로 생각하고, 반대로 중심부에서 일을 시작한다면 도시를 재건시키고 재활성화 시킬 수 있는 방면으로 생각해야 한다.

도시를 존재감을 드러내지만 거대한 부재 또한 느낄 수 있는, 커다란 빈 공간을 만들어내는 기념물로 생각해야 한다. 기념비적인 보이드 (void). 실현시키는 현실에서 잠시 벗어나 자극적인 현실로 들어가 보아라. 도시의 새로운 발전 모델과 이미지들을 찾아봐라. 이미지를 통해 새로운 모델과 패러다임을 생각해보아라. 하지만 어떠한 이미지들과 모델들을 사용해야 하나?

오토리드는 이렇게 말했다. "이미지 없이 기억은 없고, 기억 없이 이미지는 없다. 이미지는 기억의 일부분이다."

이것은 우리가 과거를 어떻게 생각하느냐가 아닌, 미래를 어떻게 생각하느냐를 보여주는 것이다. 우리가 어떻게 상상하는지를 나타내는 것이다. 이미지란 무엇인가? 본질에서 벗어나 독립적으로 존재할 수 있는 것인가? 들뢰즈는 '생각 속에 생각'이란 이미지 쇼의 결과물이라고 했다. 아이디어가 뚜렷하지 않아도 이미지는 항상 뚜렷할 것이다.

슈퍼 보부르는 건물들이 상징적인 것이 아니라 실현되는 현대성의 새로운 논리를 나타낸다.

구조물 같은 건물들, 나열하기도 힘들다.

보드리야르는 보부르를 이렇게 정의한다. "흡수하고 자라나는 매스-컬처, 이 거대한 문화 디바이스는 내부와 외부의 관계를 해결하고 도시의 문화와 역사를 흡수하고 재해석한다. 자신이 변하면서 세상에 알리는 역사다."

새로운 현대성 논리에 의하면 이러한 건물들은 자라면서 모든 것을 포함시킨다. 도시가 커지면서 재활용되지만, 새로운 지역으로 넓혀가는 것이 아니라 자기 자신을 포화시키며 소비시킨다.

렘 콜하스: 도시는 자신을 재사용한다!

오스발트 마티아스 웅어스는 이렇게 얘기했다. "그의 건물은 로멘틱하지도 이성적이지도 전통적이지도 현대적이지도 않다. 하지만 그 장소의 역사와 현실, 그리고 그곳에서 사는 사람들의 현실과 강하게 연결하고자 한다."

쇼펜하우어는 "컨셉은 죽은 탱크와도 같다. 안에 넣어진 것들은 모두 함께 있지만, 안에 넣은 것보다 더 꺼내는 것은 불가능하다. 대신에 아이디어는 같은 컨셉에 비교했을 때 완전히 새로운 것을 표현하고 발전시킬 수 있다. 제한되지 않은 것을 만들어내는 살아있는 유기체와도 같다.

OMU는 이렇게 말했다. 멘탈 이미지 없이 생각하는 것은 불가능하다.

건축은 완전히 자기 지식적이다. 자신이 알아서 지식을 만들어낸다. 건축가들은 표현과 지식의 장으로 더 이상의 도시들을 필요로 하지 않는다. 하지만 미디어는 필요로 한다. 이것은 바우만의 유동하는 세계처럼 건축을 접할 수 있는 활동 범위 표현법을 넓혀줬지만, 자율적인 것으로부터 멀어지게 하여 건축법과 역사에 의해 규제되도록 하였다. 전에는 건축의 사회적 가치가 이곳에 가장 큰 비중을 두고 있었지만, 이제는 그 공간이 어떻게 사용되고 있는가에만 그 가치가 있다.

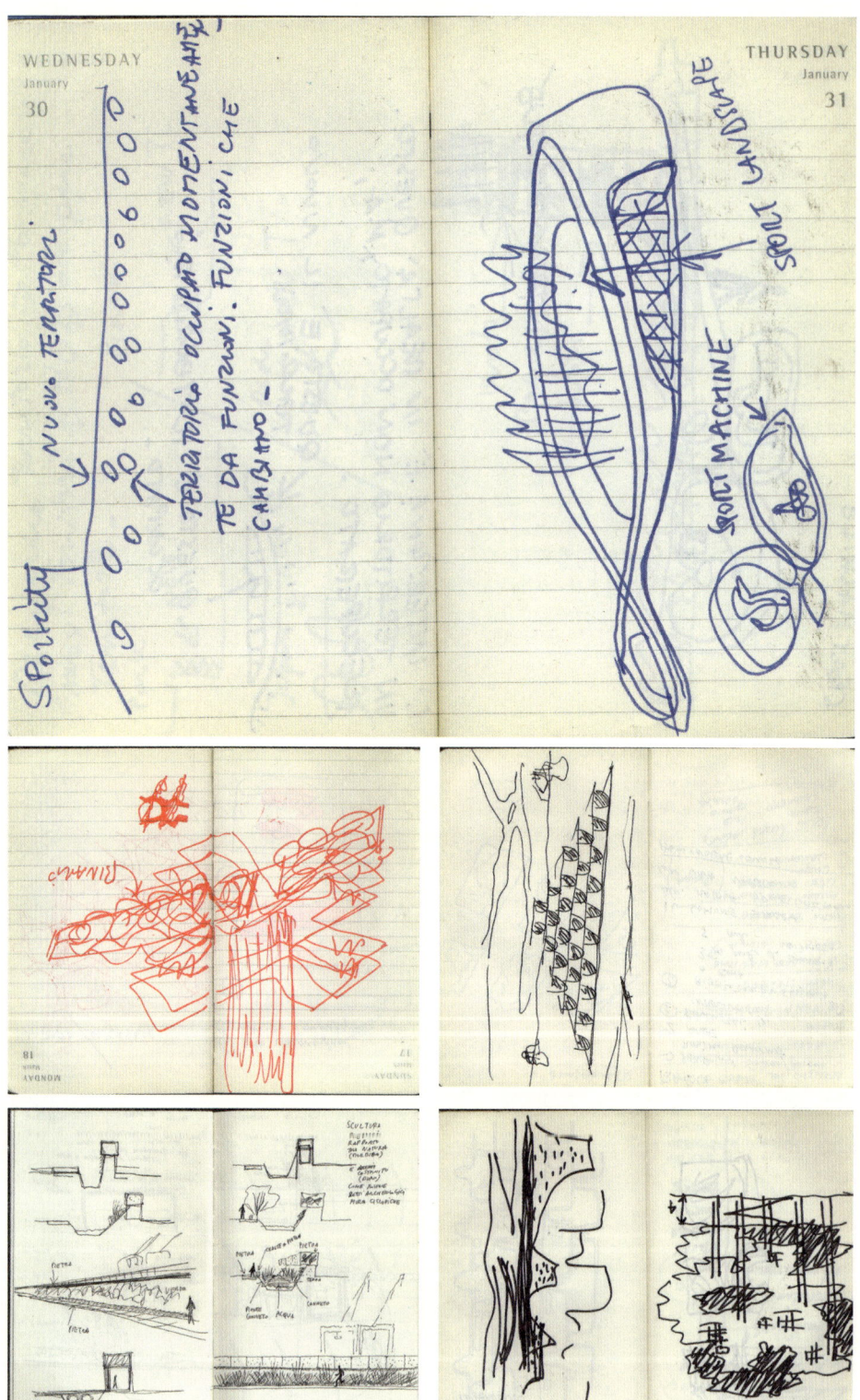

14

-Collages

LIMA URBAN LAB

www.lul-lab.com

Lima Urban Lab combines the academic and research work with the professional practice developing architecture and design projects such as: Jose de San Martin Public High school, Pisco, Peru (2007-2010), Fabula bakery and pastry boutique (2007-2009), High school Pio XII Cafeteria, Lima (2007-2010), and Sports Complex for the Jose de San Martin Public High school (2010-), Swimming Pool Santa Ana Public High school (2010-). Currently under development three Public High schools: Mateo Pumacahua High School (Cusco-Peru), Argentina High School (Chimbote, Peru) and Agustin Gamarra High school (Cusco-Peru).

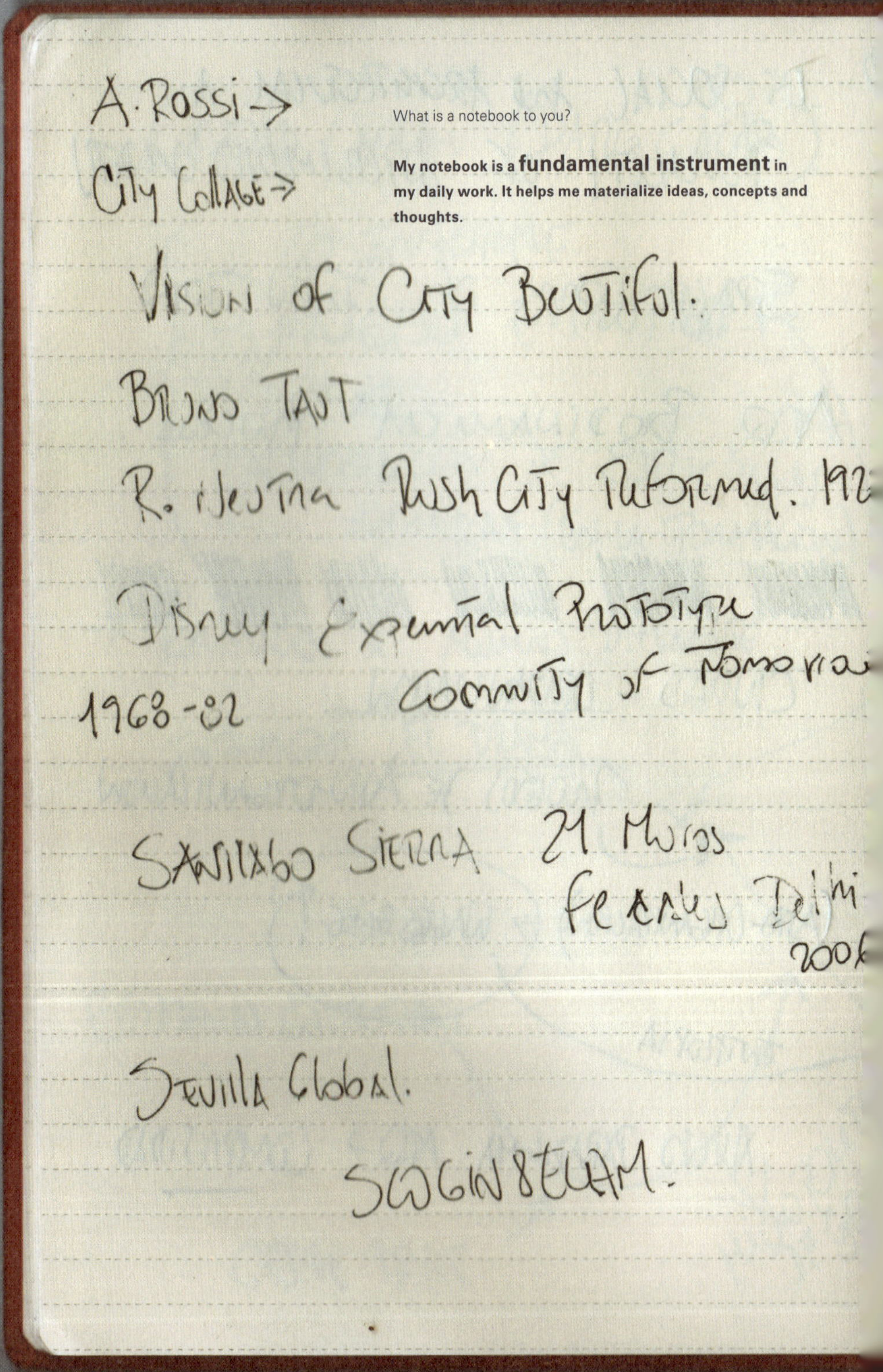

A. Rossi →
City Collage →

> What is a notebook to you?
>
> **My notebook is a fundamental instrument** in my daily work. It helps me materialize ideas, concepts and thoughts.

Vision of City Beutiful.

Bruno Taut

R. Neutra Rush City Reformed. 192

Disney Experimental Prototype
            Community of Tomorrow
1968-82

Santiago Sierra 21 Muros
                 Fe crkes Delhi
                            2006

Sevilla Global.
        SWGIN 8 ELAM.

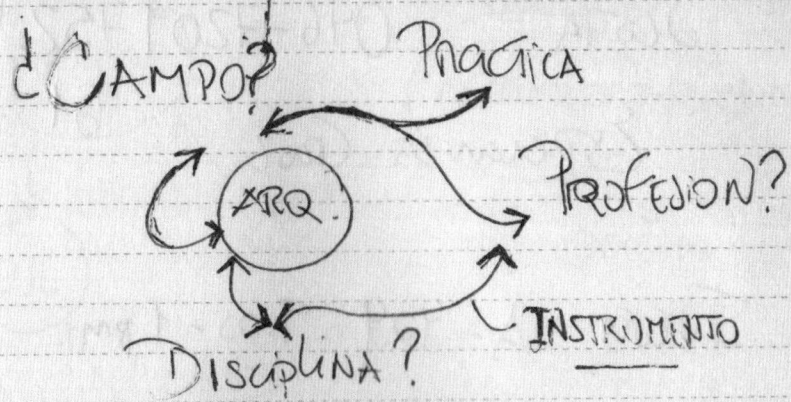

¿CAMPO? PRACTICA
ARQ
PROFESION?
DISCIPLINA? INSTRUMENTO

— NEGOCIAR LA REALIDAD —

- QUE ES SER ARQUITECTO
- FORMACION Y ~~PROFESION~~
  PROFESION
    - ADQUIRIR UNA SENSIBILIDAD ESPECIAL
    - BUSQUEDA PERFECCION

- CREATIVO / CONSTRUCCION / URBANISTA

- FORMACION CONTINUA.

당신에게 수첩이란 무엇인가?

나의 수첩은 나의 일과를 기록할 수 있는 **기본적인 도구**이다.
아이디어나 컨셉, 그리고 생각들을 구체화할 수 있도록 도와준다.

Any episodes or memories related to a notebook?

I remember at the beginning of my practice that I used to make my own notebook. I bought some white sheets and cardboard.

When and where do you use your notebook the most?

I carry my sketch all the time.

수첩에 관련된 에피소드가 있다면 들려달라.

사무실을 처음 시작했을 때에 내가 나의 노트북을 직접 만들었던 기억이 난다. 흰 종이 몇 장과 판지를 사서 만들었었다.

수첩을 가장 많이 사용하는 공간과 때는?

항상 지니고 다닌다.

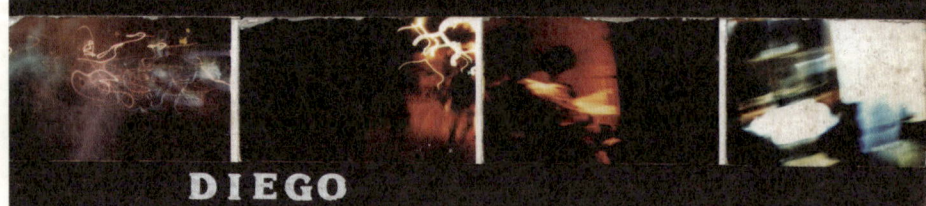

DIEGO

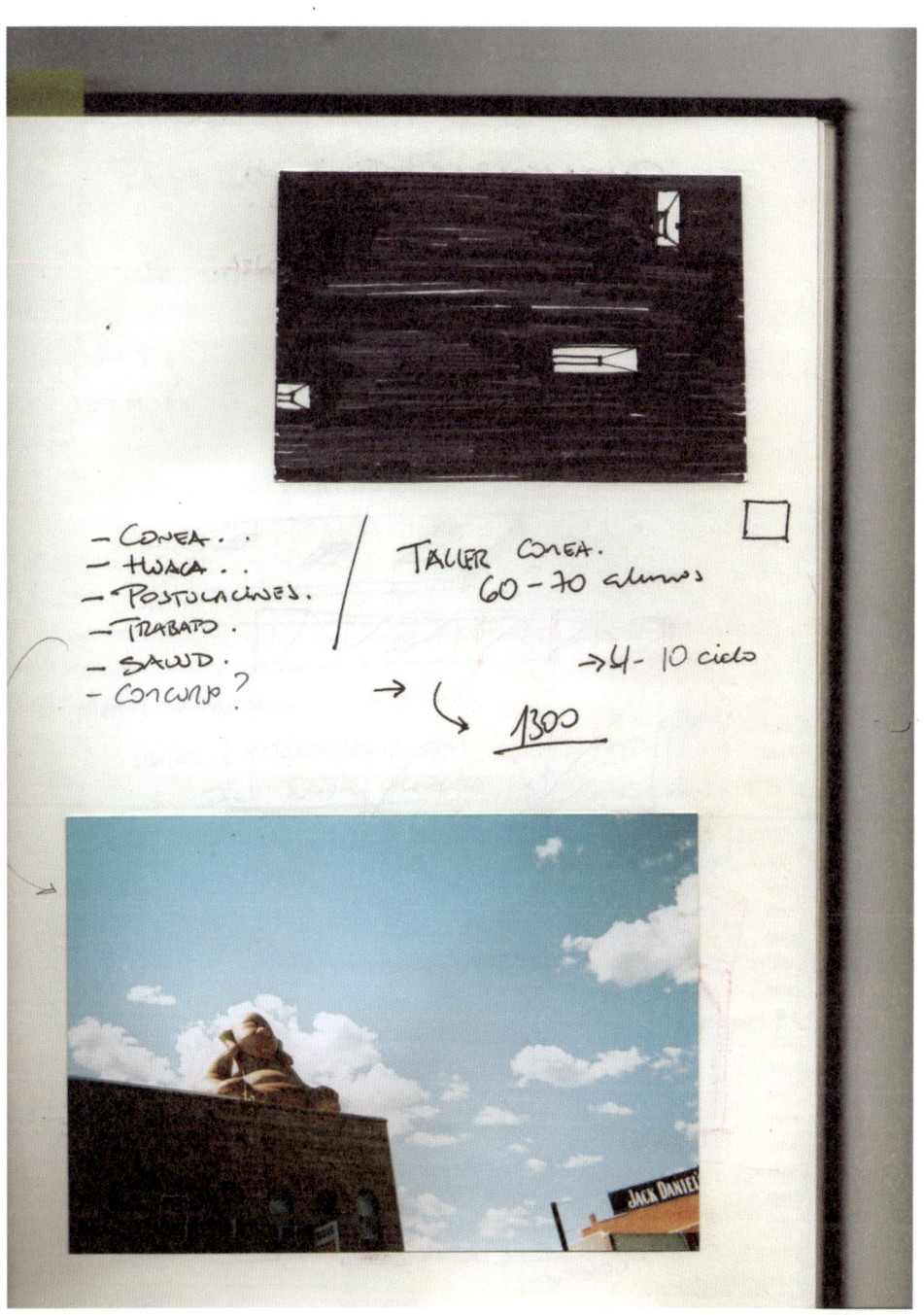

- CONEA..
- HUACA..
- POSTULACIONES.
- TRABAJO.
- SALUD.
- CONCURSO?

TALLER CONEA.
60 - 70 alumnos

→ 4 - 10 ciclo

→ ↳ 1300

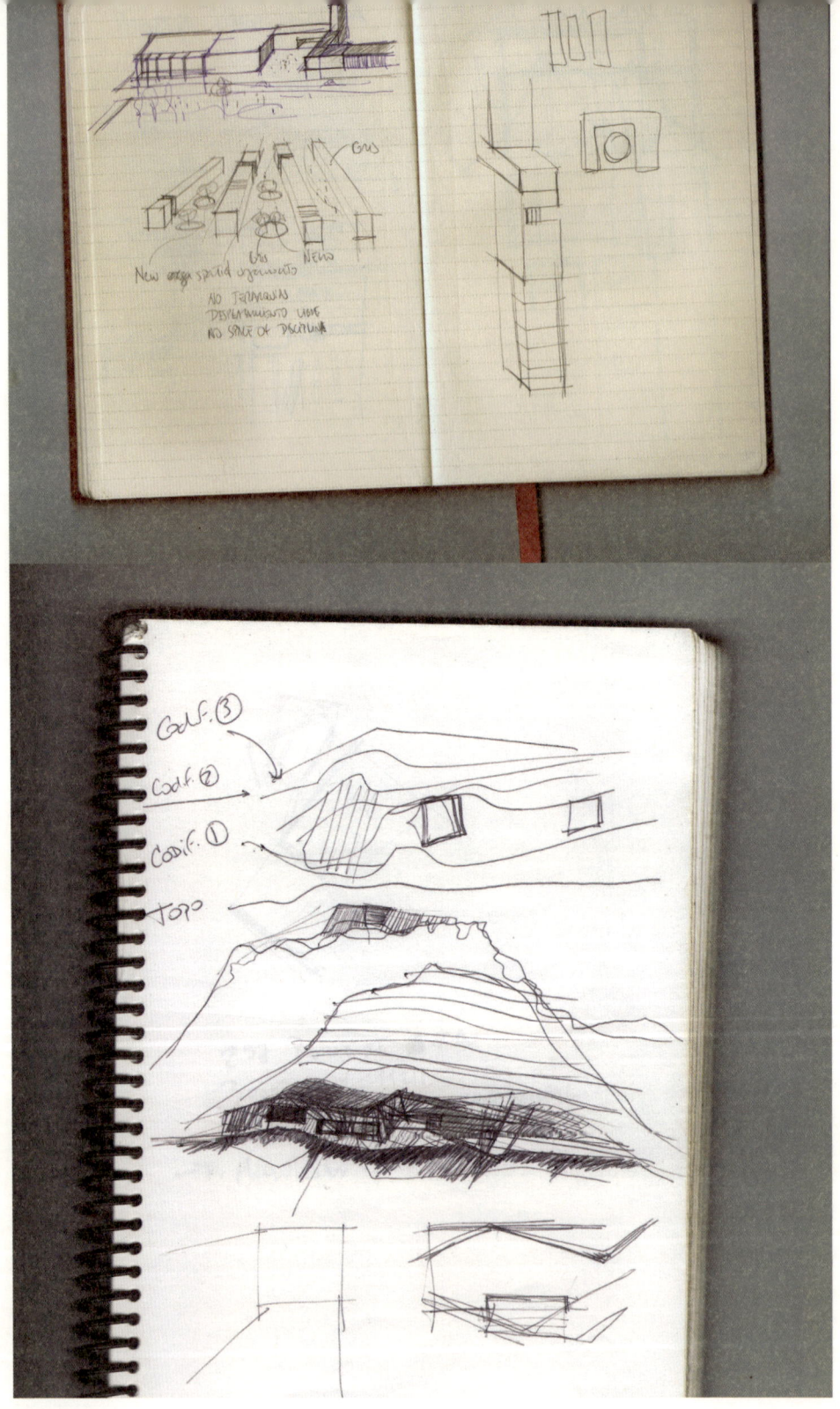

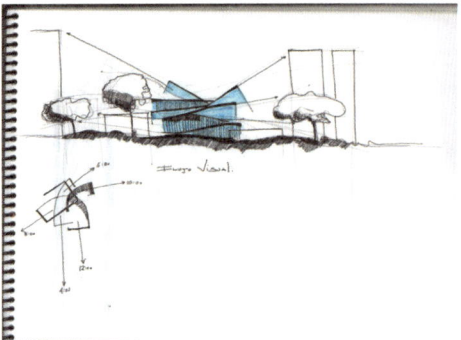

What influence does a notebook have in your projects and life as an architect?

An architect communicates through graphic expression. This is why the notebook helps us to always exercise our ideas with diagrams, mental maps, drawings, etc.

수첩은 당신의 작품과 삶에 어떤 영향을 끼치는가?

건축가는 그래픽 표현을 통해 소통하기 때문에 수첩이 중요하다. 아이디어를 다이어그램, 인식도, 도면 등등으로 표현하는 법을 연습할 수 있게 해주기 때문이다.

Are there anything else other than a sketchbook that you use to keep a record of your thoughts and ideas?

I use my notebook most of the time.

수첩 외에 자신의 생각을 기록하는 방법과 도구는 무엇이 있는가?

나는 거의 항상 수첩을 사용한다.

LIMA – TACNA
TACNA – ANTOFAGASTA
CASA EN EL PANTANO
Nivel 2½

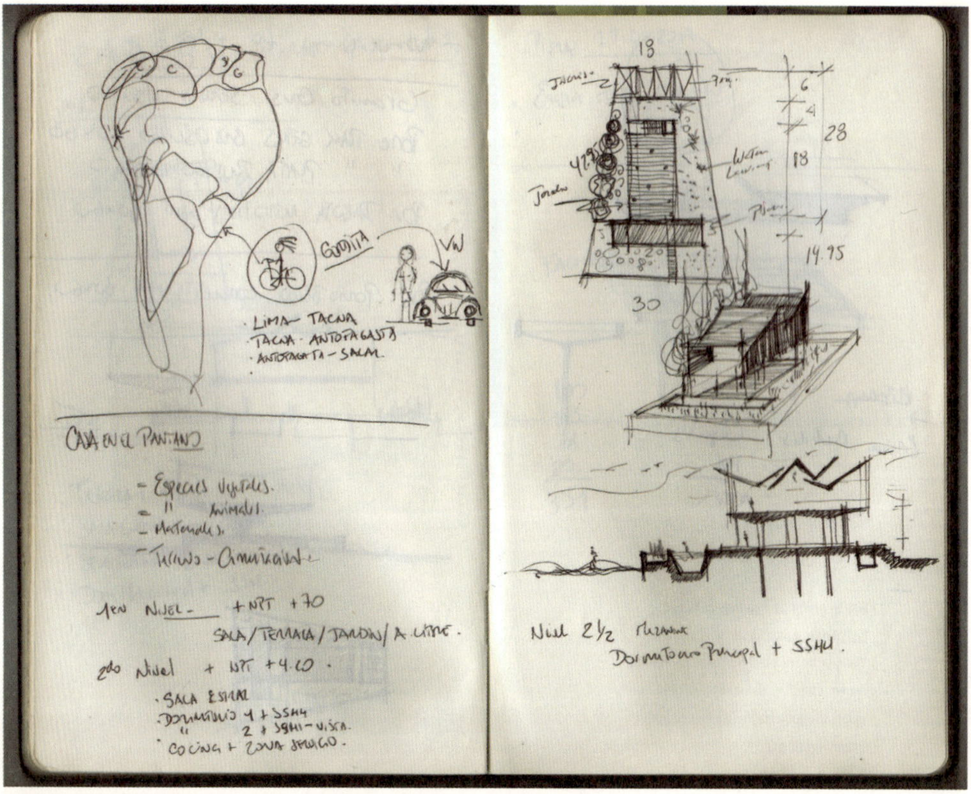

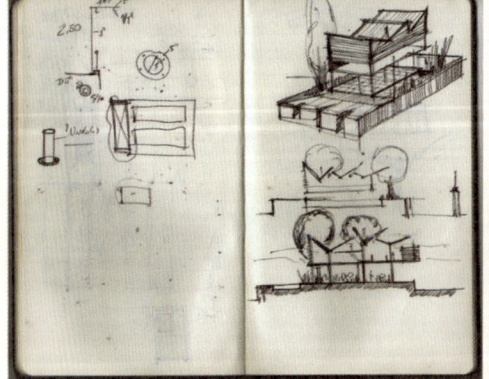

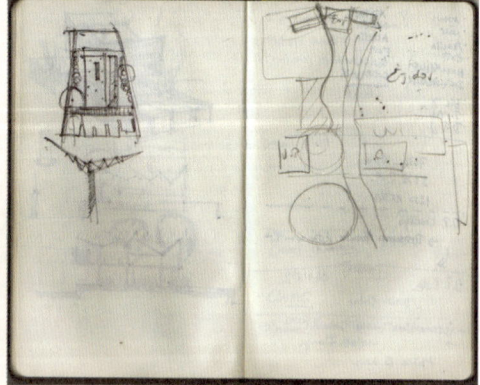

*Country House

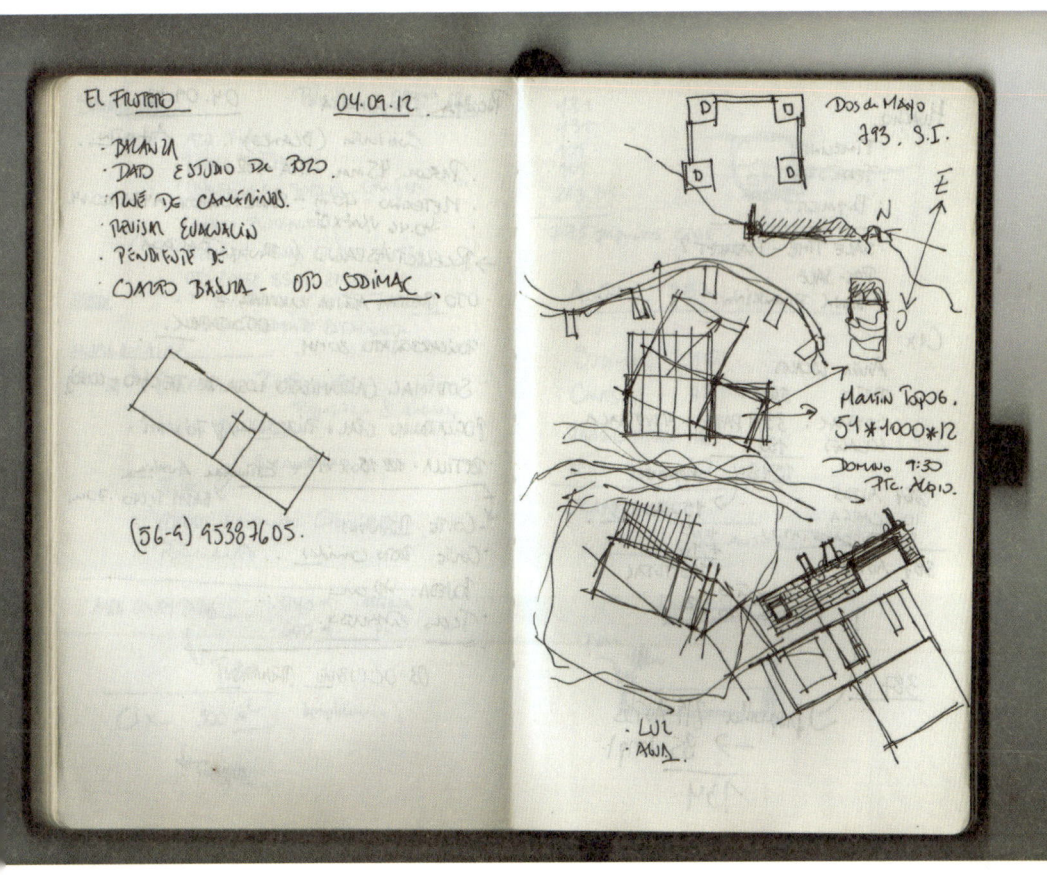

I remember at the beginning of my practice that I used to make my own notebook. I bought some white sheets and cardboard.

Sport facilities jose de san martin highschool

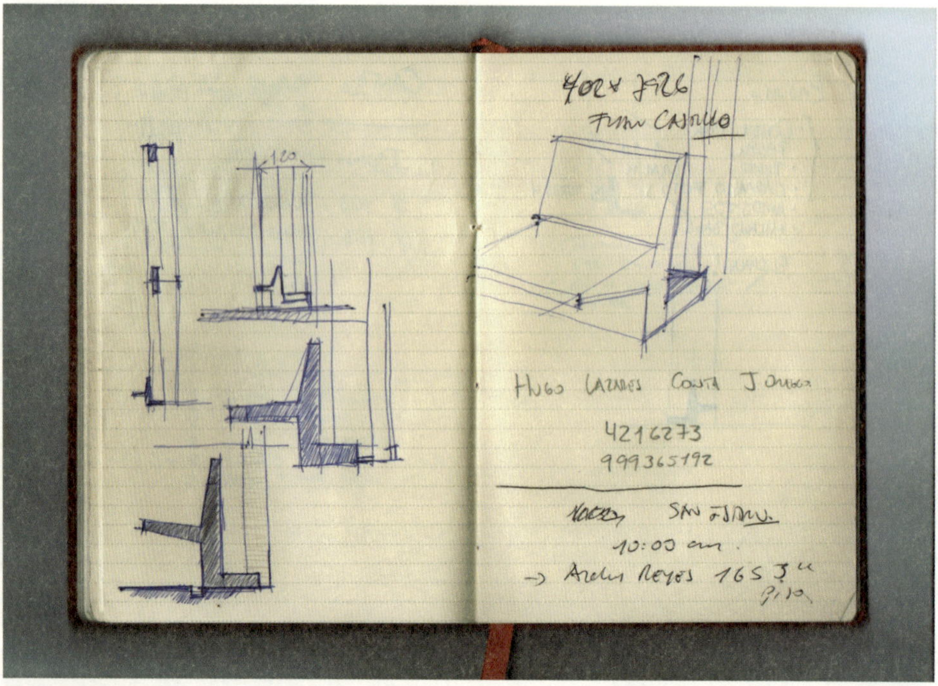

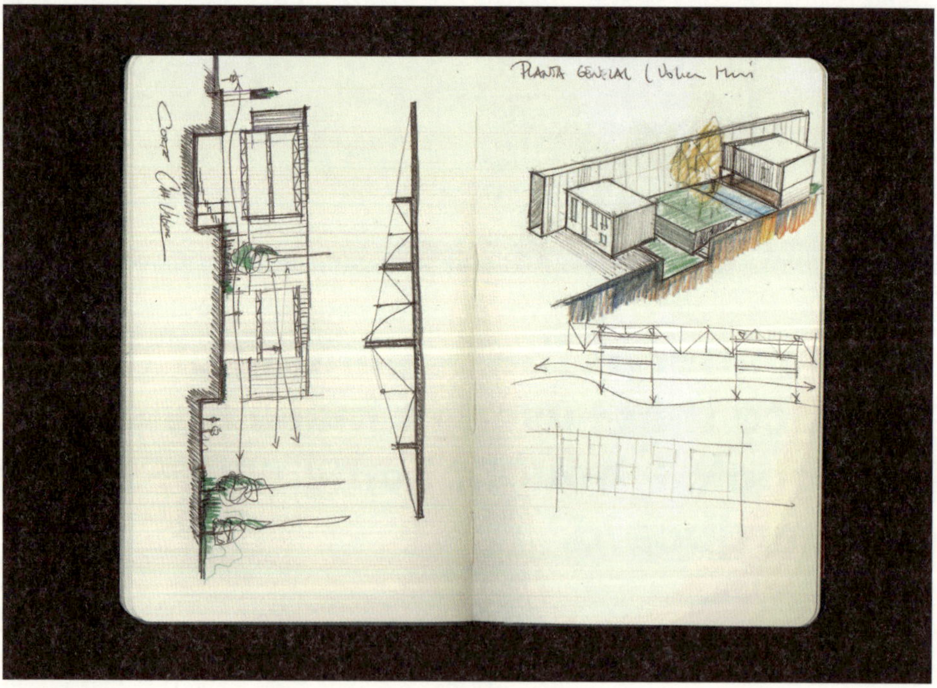

Urban House

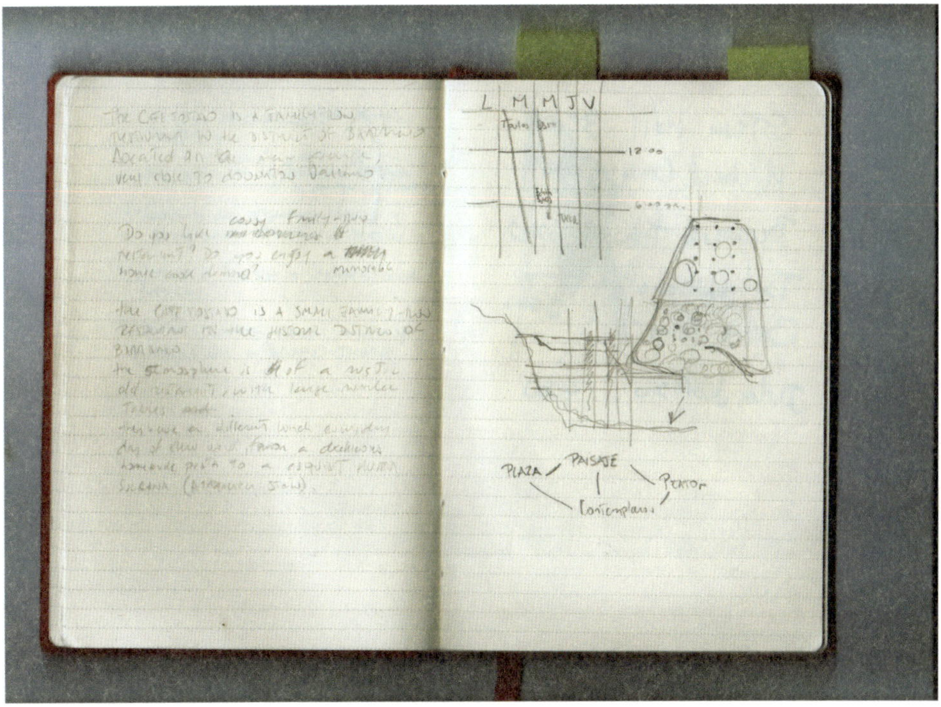

Memory Museum

ZO_LOFT
ARCHITECTURE & DESIGN

www.zo-loft.com

ZO_loft architecture & design was born, offering a mobile, sustainable, and flexible architecture, design and urban space.
Since its founding ZO_loft is networking with national and international partners, with the addiction of dissimilar competence and personality mixed with an ongoing research and experiences.
ZO_loft is an architectural and product design office working on architecture and urban planning; exhibition and interior design, product and industrial design, product development, graphics, technical consulting and art direction for companies and events.

Three consecutive years finalist at Macef Design Award, ZO_loft also won several international awards, among which: First Prize Macef Design Award with the project Din-ink. In May 2012 they got two honorable mentions for MA Prize Atlanta. In 2009 ZO_loft got a Nomination for Well Tech International Award with the project Wheely and had the honor to be called to represent its Country at "Italia in Giappone 09" with the project WinOt.
In 2010 ZO_loft was included in the annual of Young Italians talent in the World YOUNG BLOOD.

What is a notebook to you?

Once Walter Benjamin said *"Do not miss any thought, and keep your notebook as the authorities keep a register of foreigners"*. Well Actually it's like that! A sketchbook it is like a partner or a collegue that helps you reminding your ideas, or **to preserve an emotion before it becomes too "reasoned"**. But it's also like meeting an old friend after a long time. Looking through it you will remember a lot of things you had probably forgot about your life.

당신에게 수첩이란 무엇인가?

발터 벤야민이 이런 말을 했었다. "어떠한 생각도 놓치지 마라. 그리고 당국이 외국인 등록부를 관리하듯이 너의 수첩을 가지고 있어라". 정확히 이렇다! 스케치북은 아이디어를 상기시켜주는 파트너나 동료와도 같은 것이다. 또는 **감정이 너무 '이성적'이게 되지 않도록 해주는 것이다.** 하지만 오랜 시간이 흐른 후 옛 친구를 만나는 것과도 같다. 스케치북을 보면서 삶에서 잊어버렸던 많은 것들을 다시 기억하게 될 것이다.

359

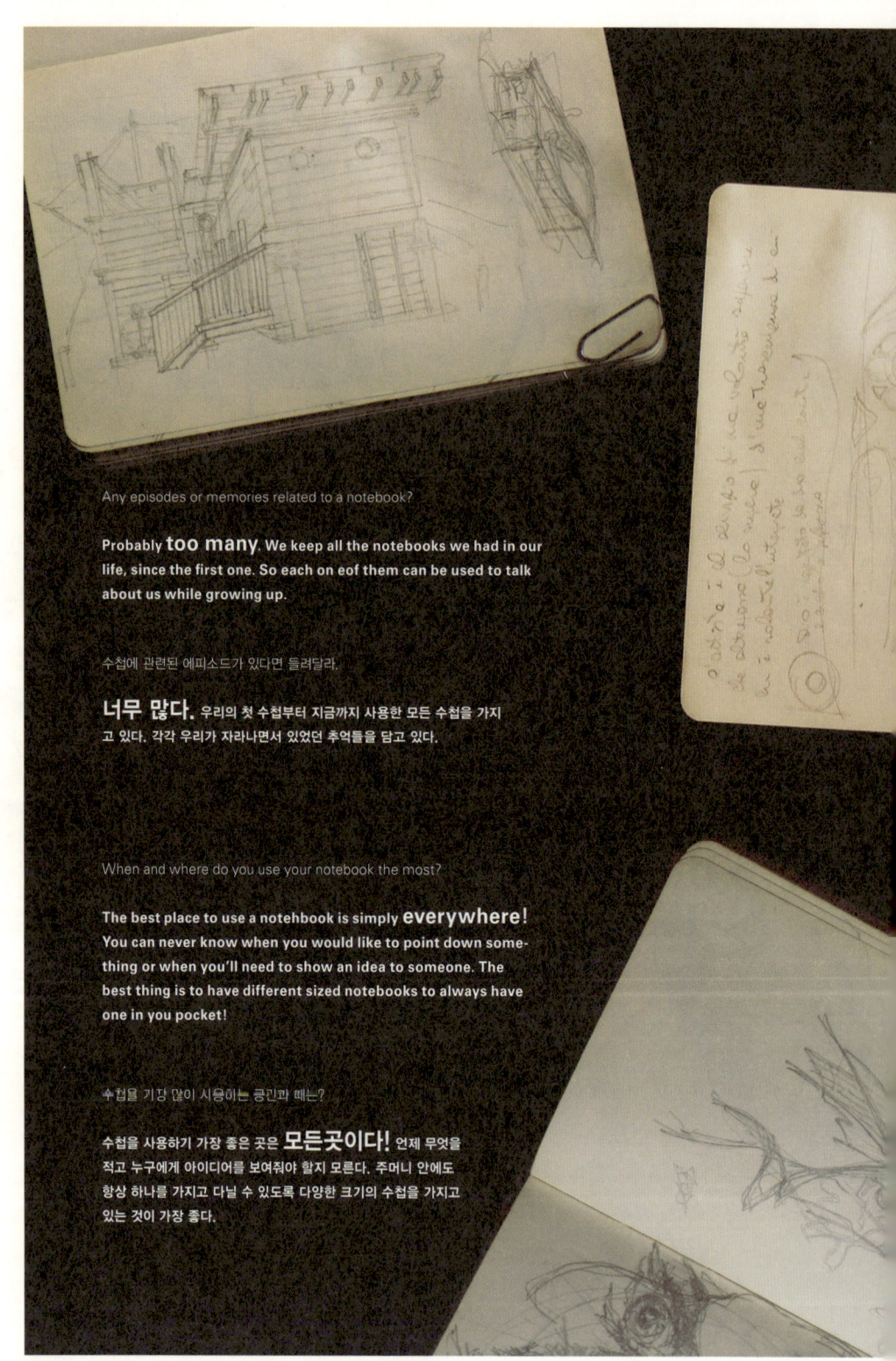

Any episodes or memories related to a notebook?

Probably too many. We keep all the notebooks we had in our life, since the first one. So each on eof them can be used to talk about us while growing up.

수첩에 관련된 에피소드가 있다면 들려달라.

너무 많다. 우리의 첫 수첩부터 지금까지 사용한 모든 수첩을 가지고 있다. 각각 우리가 자라나면서 있었던 추억들을 담고 있다.

When and where do you use your notebook the most?

The best place to use a notehbook is simply **everywhere!** You can never know when you would like to point down something or when you'll need to show an idea to someone. The best thing is to have different sized notebooks to always have one in you pocket!

수첩을 가장 많이 사용하는 공간과 때는?

수첩을 사용하기 가장 좋은 곳은 **모든곳이다!** 언제 무엇을 적고 누구에게 아이디어를 보여줘야 할지 모른다. 주머니 안에도 항상 하나를 가지고 다닐 수 있도록 다양한 크기의 수첩을 가지고 있는 것이 가장 좋다.

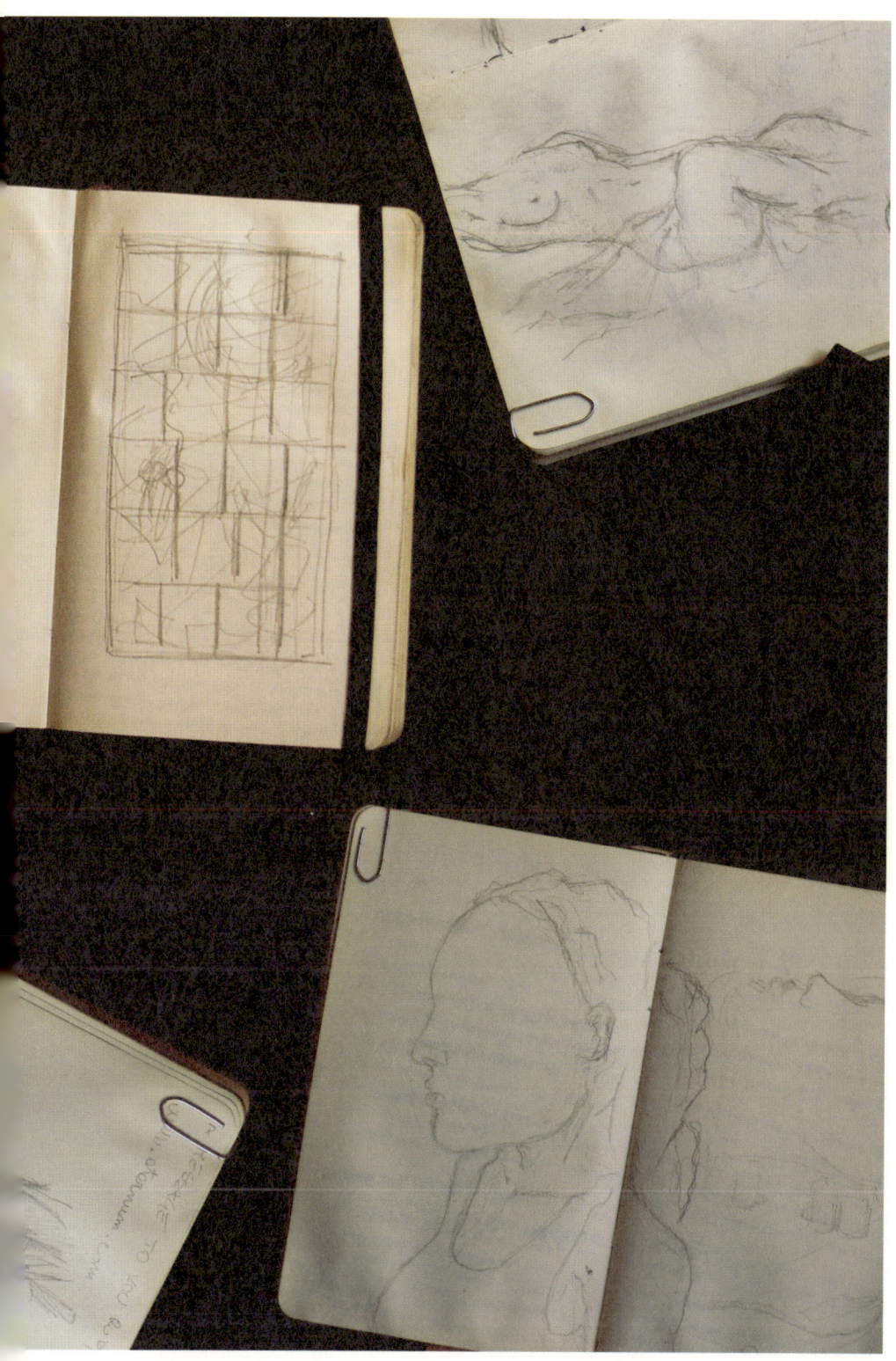

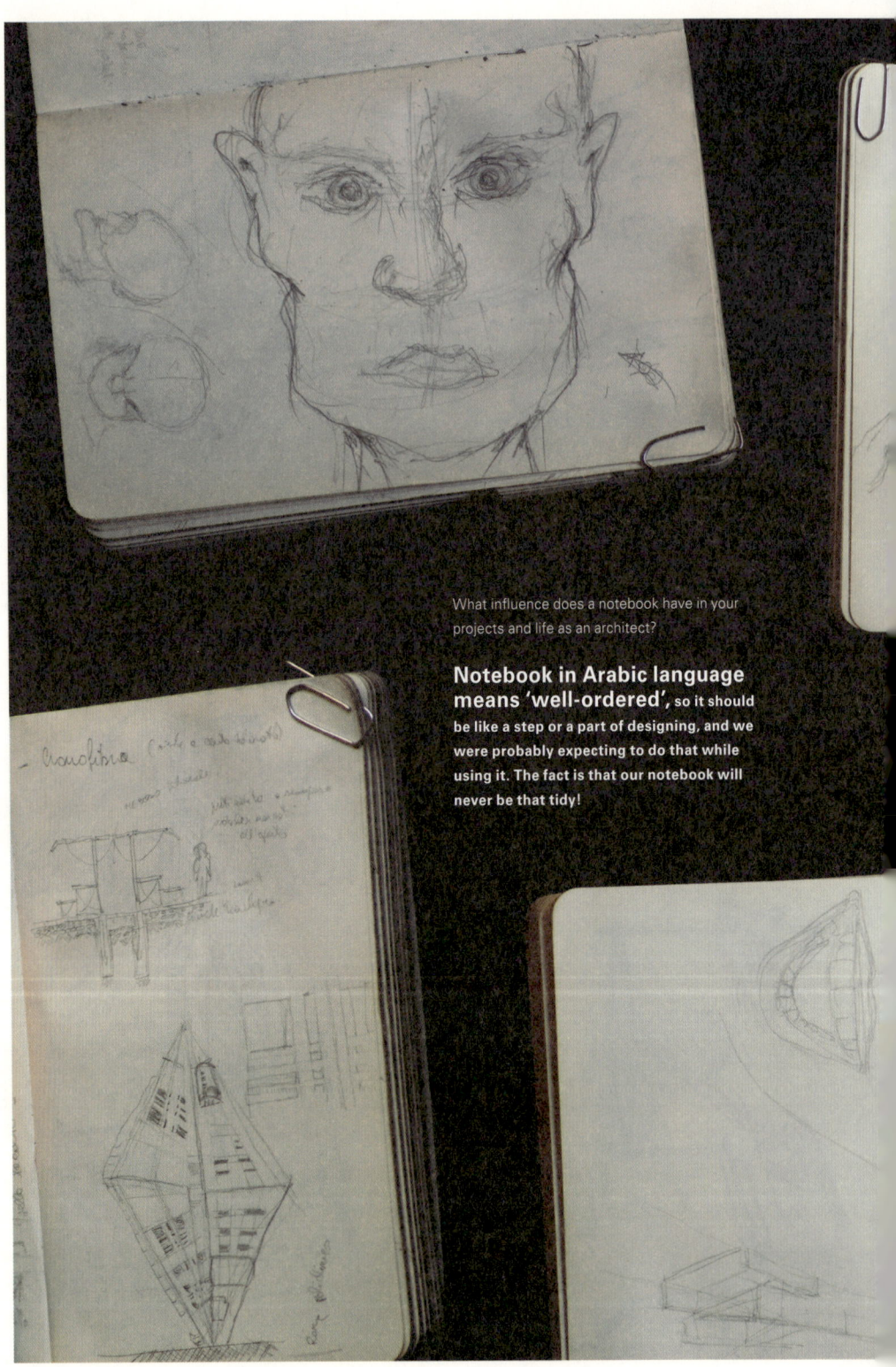

What influence does a notebook have in your projects and life as an architect?

Notebook in Arabic language means 'well-ordered', so it should be like a step or a part of designing, and we were probably expecting to do that while using it. The fact is that our notebook will never be that tidy!

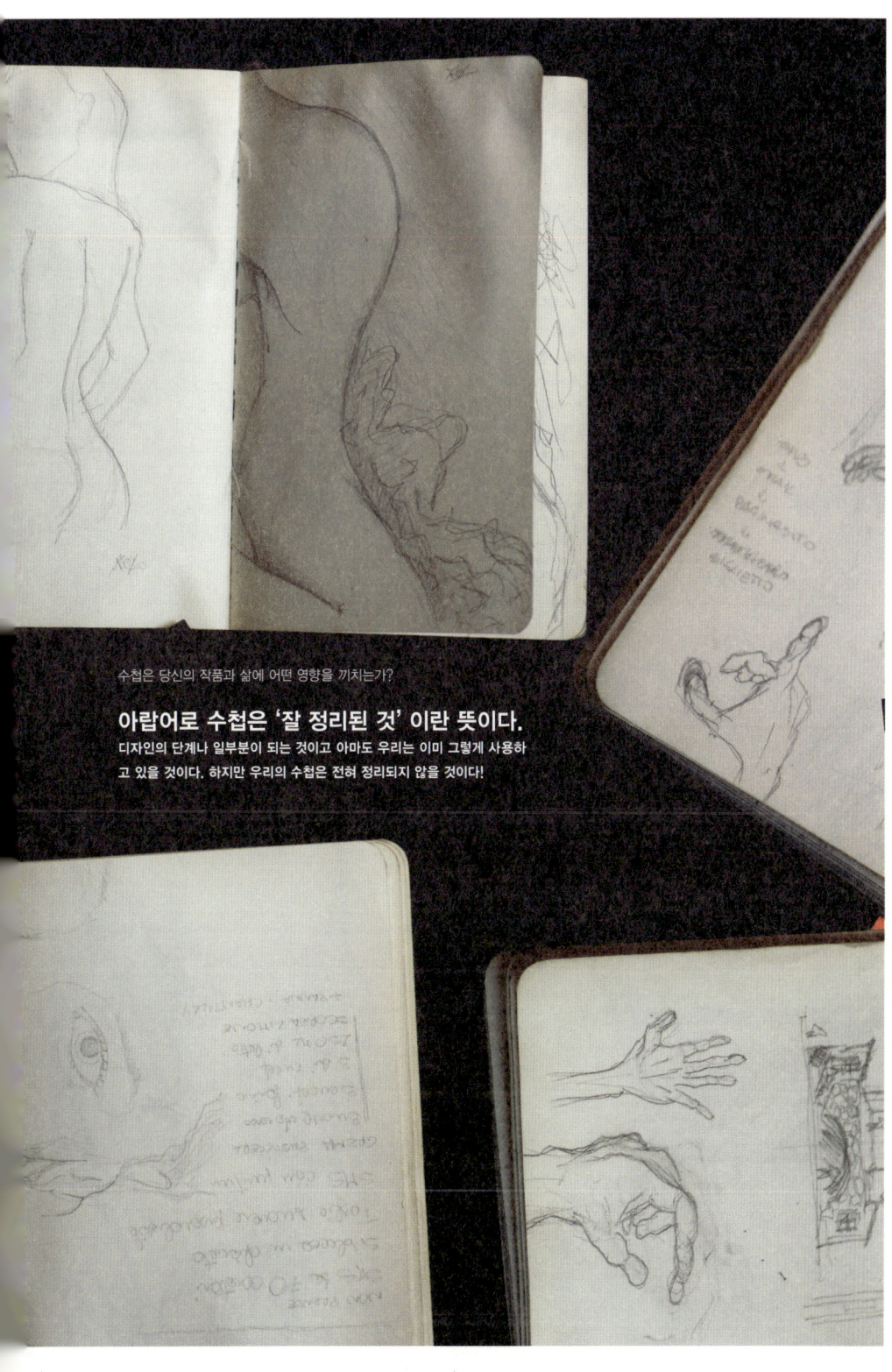

수첩은 당신의 작품과 삶에 어떤 영향을 끼치는가?

아랍어로 수첩은 '잘 정리된 것' 이란 뜻이다.
디자인의 단계나 일부분이 되는 것이고 아마도 우리는 이미 그렇게 사용하고 있을 것이다. 하지만 우리의 수첩은 전혀 정리되지 않을 것이다!

Notebook in Arabic language means 'well-ordered'.

Are there anything else other than a notebook that you use to keep a record of your thoughts and ideas?

Everything you can write on. For real. The problem is that sometimes you need to copy them because you can not take the table or the wall with you!

수첩 외에 자신의 생각을 기록하는 방법과 도구는 무엇이 있는가?

그릴 수 있는 것은 **무엇이든지** 사용한다. 사실이다. 문제는 가끔은 그린 것을 복사해야 할 때가 있다는 것이다. 테이블이나 벽을 가지고 갈 수는 없으니 말이다!

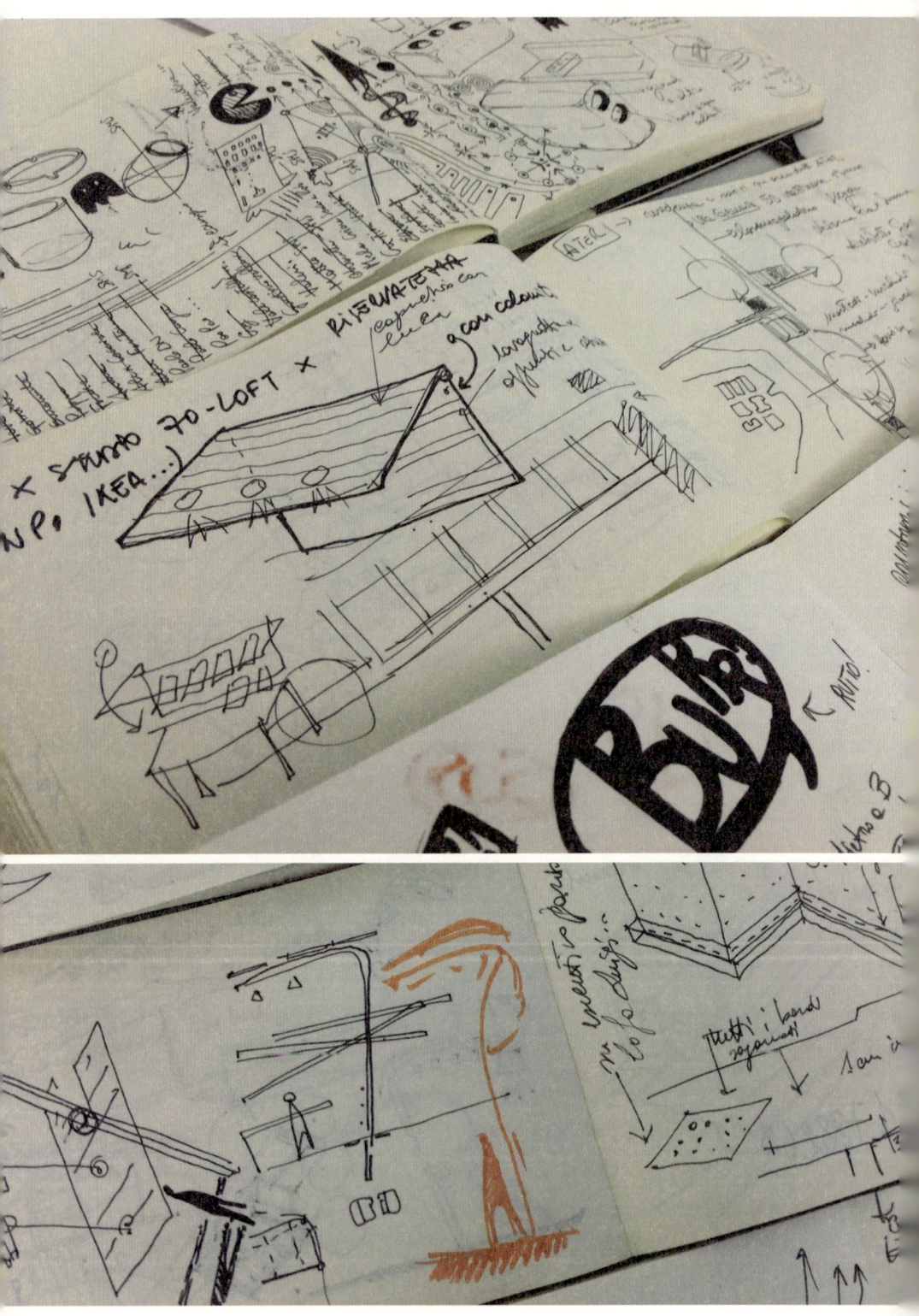

IAD PARTNERS

www.groupiad.com

Stéphane Cottroll, Partner & Administrator at IAD Independent Architecture Diplomacy, Paris-La-Seine School of Architecture, 1994.
Of the 3 partners, Stéphane, in charge of the international development of the studio, is the one who travels the most. He's unconditionally linked to his notebooks, travelogues, of ideas and sketches that allow him to stay in contact with the studio even on the other side of the world, and to be more involved in the creative process by preserving in his notebooks, thoughts that may disappear in the significant amount of data brought from his faraway destinations.

What is a notebook to you?

If my notebook were computerized, it would be an external hard drive, a memory asset of my thoughts concerning projects and diverse architectural situations. But it's a paper pad that must be a "Moleskine". Regardless of the size or thickness of it, I have an exclusive relation with this reference. If I happen to forget it, a piece of paper tablecloth, a bag or a napkin will do. Each of these drawings will end in my Moleskine notebook anyway.

Any episodes or memories related to a notebook?

I have sometimes thought on delivering a whole project in sketch format. I think it would be such a time saving as the essence of these projects can be found in my notebooks. At the end of a meeting in some far away country, I like to seat in a bench (my public office in some ways) to take over with drawings the different points of view of the meeting. Without Wi-Fi or scanner, I just send some pictures of my notebook to the studio as a first debriefing. Long live Sketchbooking!

당신에게 수첩이란 무엇인가?

만약 나의 수첩이 컴퓨터화 될 수 있다면, 프로젝트와 다양한 건축에 대한 나의 생각들을 저장해 놓는 외장하드와도 같은 것이다. 하지만 '몰스킨'이어야 하는 종이 책이다. 크기와 두께와는 상관 없이 나는 수첩과 특별한 관계를 지니고 있다. 만약 수첩이 내 손에 없다면, 식탁보나 가방, 또는 냅킨으로도 대체 가능하다. 무엇이 되었든 간에, 나의 몰스킨 수첩 안에 들어갈 테니까 말이다.

수첩에 관련된 에피소드가 있다면 들려달라.

나는 가끔 프로젝트 전체를 스케치 형식으로 해볼까라는 생각을 할 때도 있다. 프로젝트들의 주요 포인트들이 모두 수첩 안에 있기 때문에 엄청난 시간 절약이 될 것 같기 때문이다. 어느 먼 나라에 갔을 때 미팅이 끝나고 나면 어디 벤치에 앉아 (어떻게 보면 나의 열린 사무실이기도 하다) 미팅에서 했던 얘기들을 다른 시각으로 생각하고 그려보는 것을 좋아한다. 인터넷이나 스캐너 필요 없이, 수첩의 사진을 찍어 스튜디오로 보내놓는다. 수첩이여 영원하라!

When and where do you use your notebook the most?

Out of the office, where the size of the notebook and his practical size make perfect sense.

What influence does a notebook have in your projects and life as an architect?

My notebook doesn't have an influence on me, but I do have an influence on it... It's sometimes a power relationship between us... Especially when I end one or start a new one.

It's sometimes a power

수첩을 가장 많이 사용하는 공간과 때는?

수첩의 크기와 실용성이 완벽하게 하모니를 이루는 사무실 밖에서 주로 사용한다.

수첩은 당신의 작품과 삶에 어떤 영향을 끼치는가?

수첩이 나에게 영향을 끼치지는 않고, 반대로 내가 수첩에 영향을 끼치는 것 같다. **수첩과 나 사이에 권력 다툼 같은 것이 존재한다.** 주로 내가 하나를 다 사용해 끝내는 순간이나, 새로 시작해야 하는 순간에 일어난다.

relationship between us.

Are there anything else other than a notebook that you use to keep a record of your thoughts and ideas?

No, I have an exclusive relation with my notebooks, almost obsessive. If I don't have it close I feel the same way everyone else feels without their smartphones... I am sure you get what I mean.

수첩 외에 자신의 생각을 기록하는 방법과 도구는 무엇이 있는가?

나는 나의 수첩들과 지나칠 정도로 특별한 관계를 가지고 있기 때문에 **다른 것은 사용하지 않는다.** 다른 사람들이 스마트폰을 가지고 있지 않을 때의 느낌을 나는 수첩이 없을 때 느낀다. 내가 무슨 말을 하는지 아마 당신을 알 것이다.

*Carre Prive Shopping Mall

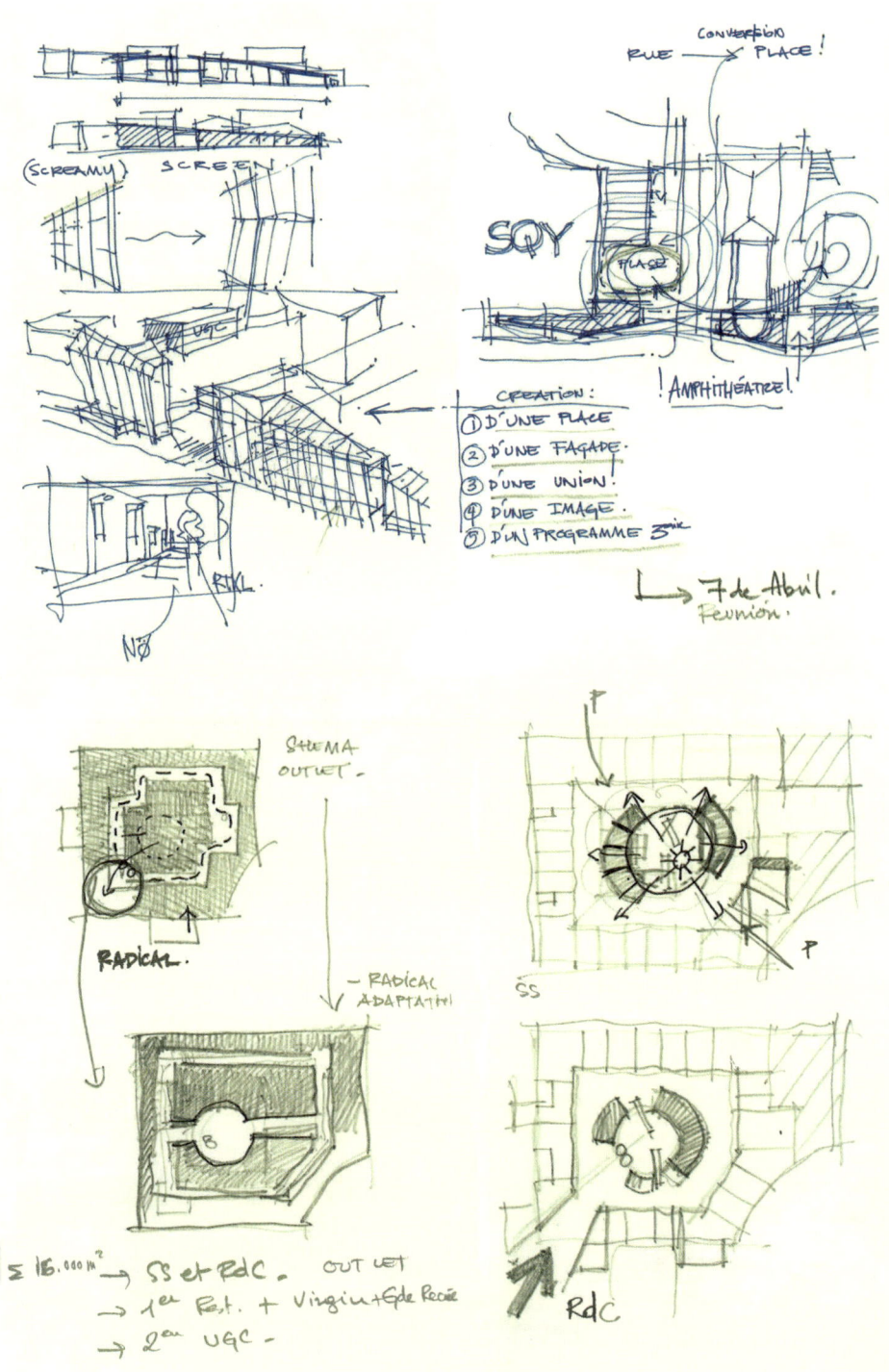

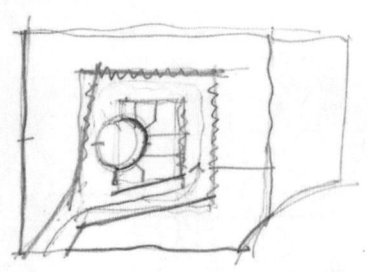
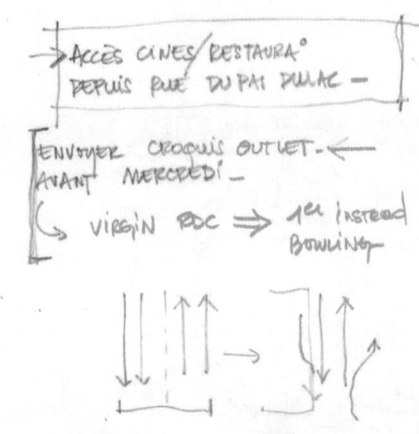

> ENVOYER A FLORENT DOSSIER –
  PLAN BY PLAN –
  HYPOTHESE MS + BOWLING SS.

> HYPOTHESE OUTLET.
  – CELLULES FERMÉES (–1/Rdc)
  1 – Gde Recta / VIRGIN Rest.
  2 – CINE.

> HYPOTHESE GD MAGASIN –
  BON MARCHE / PRIMPTEMPS.
  – DISCOUNT –

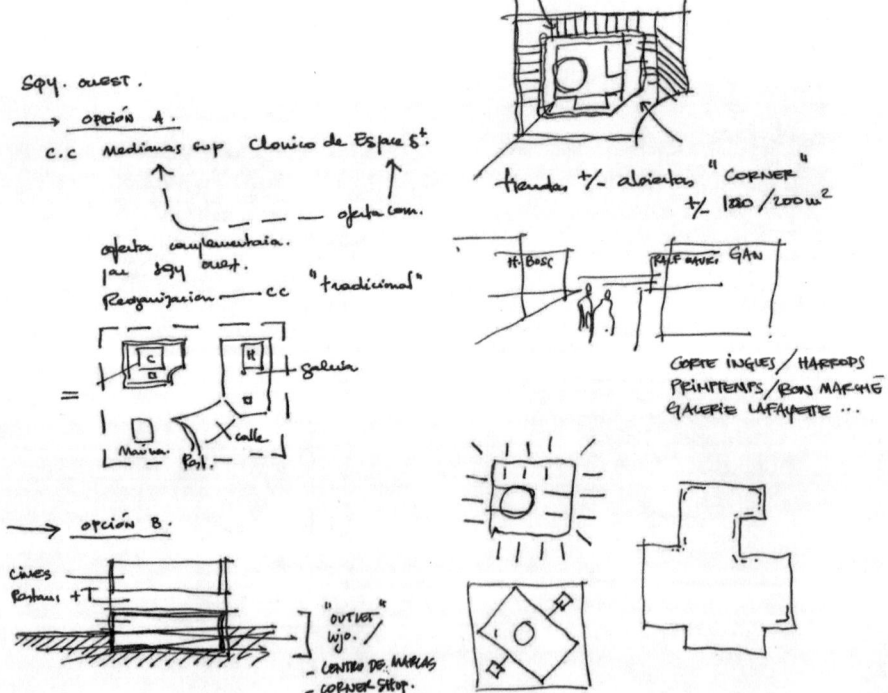

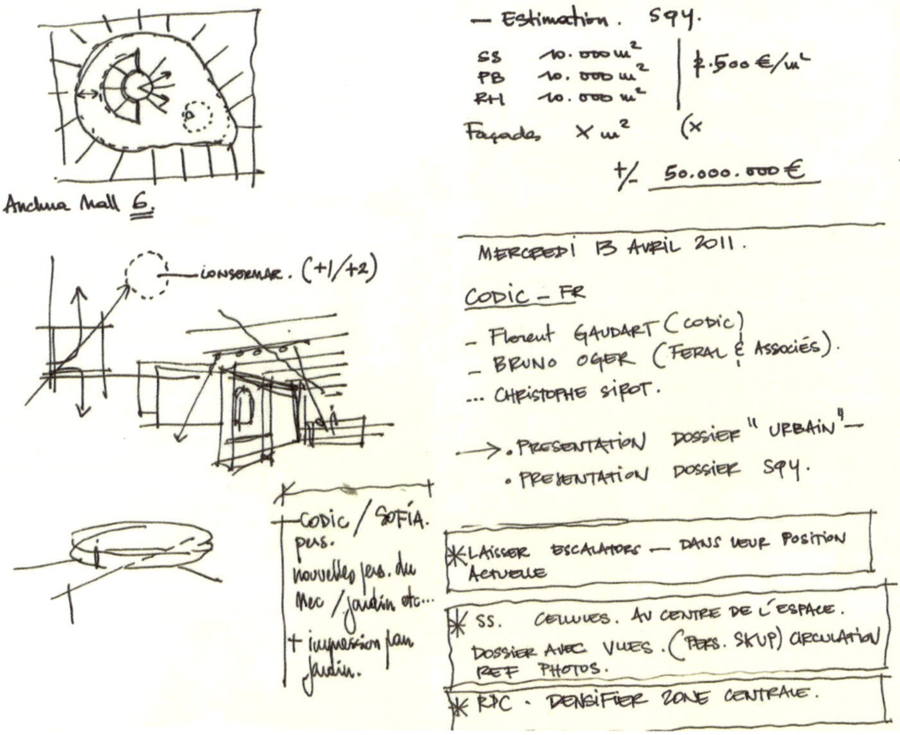

Anchura Mall 6.

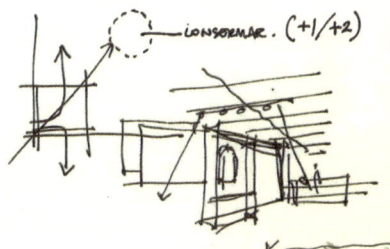

— CONSERVAR. (+1/+2)

✳ CODIC / SOFIA.
pers.
nouvelles pers. du
Mcc / Jardin etc...
+ impression pour
jardin.

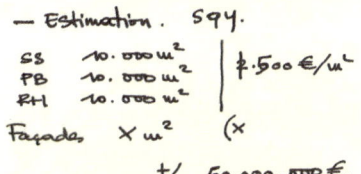

— Estimation. SPY.
SS   10.000 m²
PB   10.000 m²   } 1.500 €/m²
R+1   10.000 m²
Façade X m²   (x

+/- 50.000.000 €

MERCREDI 13 AVRIL 2011.
CODIC - FR
- Florent GAUDART (CODIC)
- BRUNO OGER (FERAL & ASSOCIÉS).
... CHRISTOPHE SIROT.

→ • PRESENTATION DOSSIER "URBAIN" —
   • PRESENTATION DOSSIER SPY.

✳ LAISSER ESCALATORS — DANS LEUR POSITION ACTUELLE

✳ SS. CELLULES. AU CENTRE DE L'ESPACE.
DOSSIER AVEC VUES. (PERS. SKUP) CIRCULATION
REF PHOTOS.

✳ RDC . DENSIFIER ZONE CENTRALE.

*Green Heart Retail Park

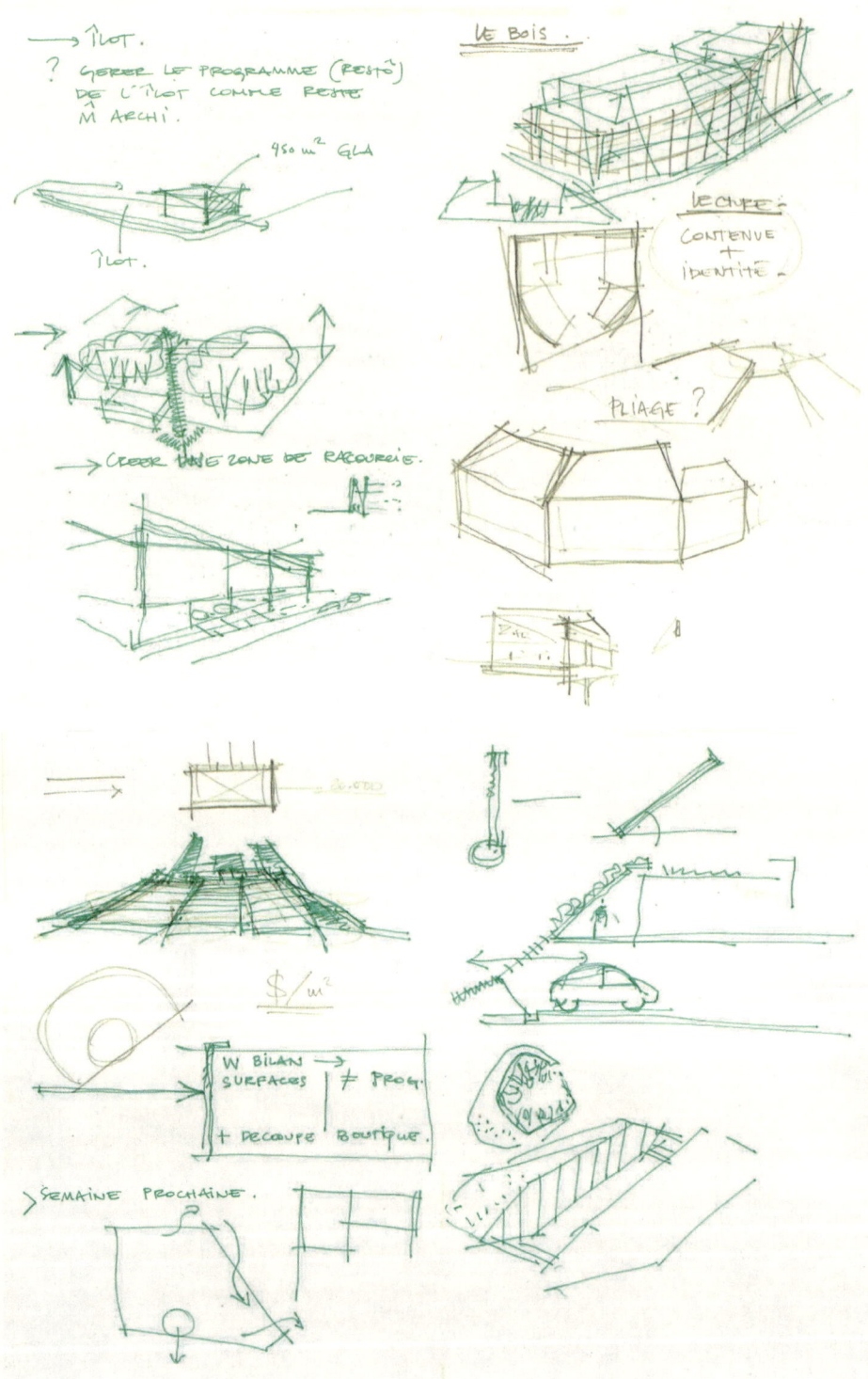

Miguel Arraiz García
[Bipolaire Arquitectos / Pink Intruder]

www.bipolaire.net / www.pinkintruder.com

Miguel Arraiz García works as an architect at his office Bipolaire Arquitectos, focusing mainly in two fields. Restoration and historical centers on one hand and sustainable architecture on the other. He is also involved in the research group Climate Symbiosis which investigates in architecture and urban planning aiming to reduce human impact. Apart from that he has a great interest in art and urban interventions, being one of the founders of Pink Intruder, where he has developed his artistic interests.

What is a notebook to you?

Is my personal diary, the place where I put my ideas during the time. Sometimes it becomes a book, because **I read it again after some time and find old ideas that were not developed at that time, but ideas that now can be used and re-developed.**

당신에게 수첩이란 무엇인가? (어떠한 의미인가?)

수첩은 언제나 내 생각들을 담아두는 개인 일기장이다. 또 가끔은 나의 책이 되기도 한다. **수첩이라는 책을 읽으며 당시에는 발전 시키지 않았던 아이디어들을 다시 꺼내어 발전시킨다.**

I read it again after some time and find old ideas that were not developed at that time, but ideas that now can be used and re-developed.

Any episodes or memories related to a notebook?

All my sketchbooks have a memory, they're always a present that someone has made me, and so they're really personal belongings.

When and where do you use your notebook the most?

Especially at home and during the night. I'm not good having only a sketchbook, I have several around the house. Sometimes it's messy because when you try to find an old idea you have to find the pieces of this idea in different sketchbooks. But it makes it more fun when you revisit your sketchbook.

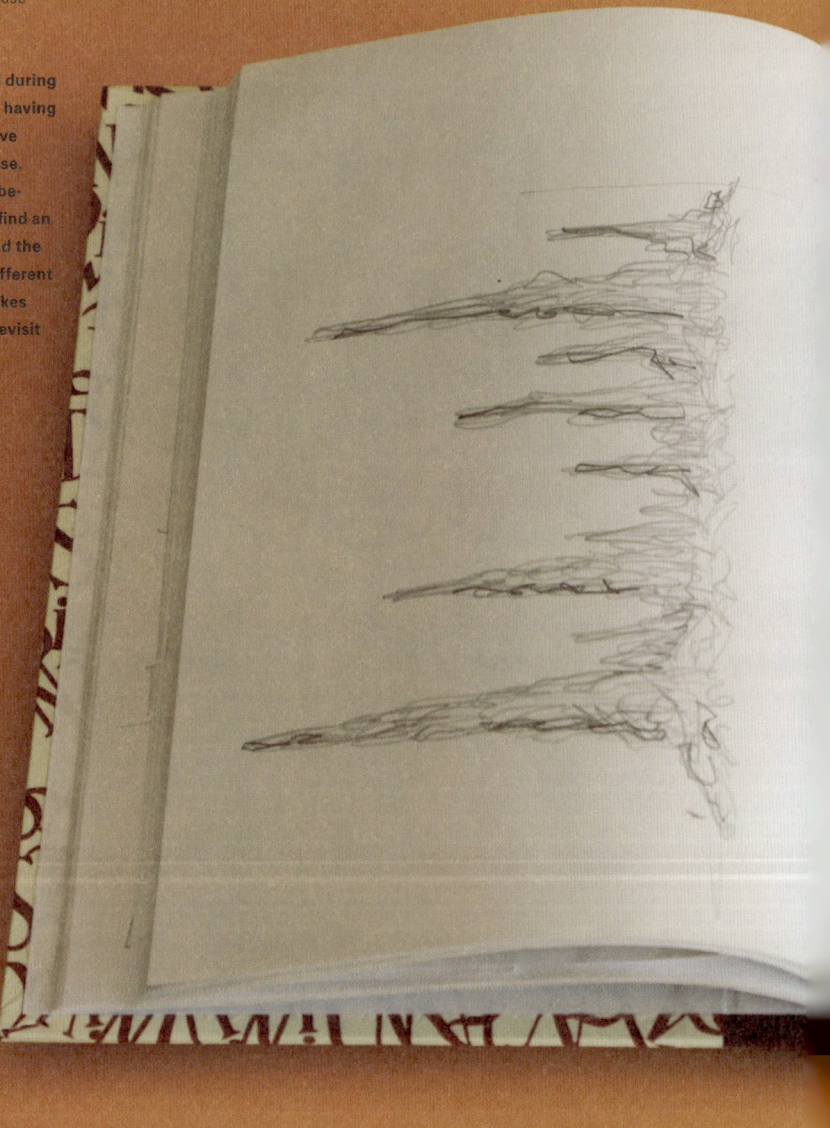

수첩에 관련된 에피소드가 있다면 들려달라.

나의 모든 수첩은 추억을 가지고 있다. 항상 누군가 나에게 준 선물이기 때문에 **하나하나 나의 소지품들이다.**

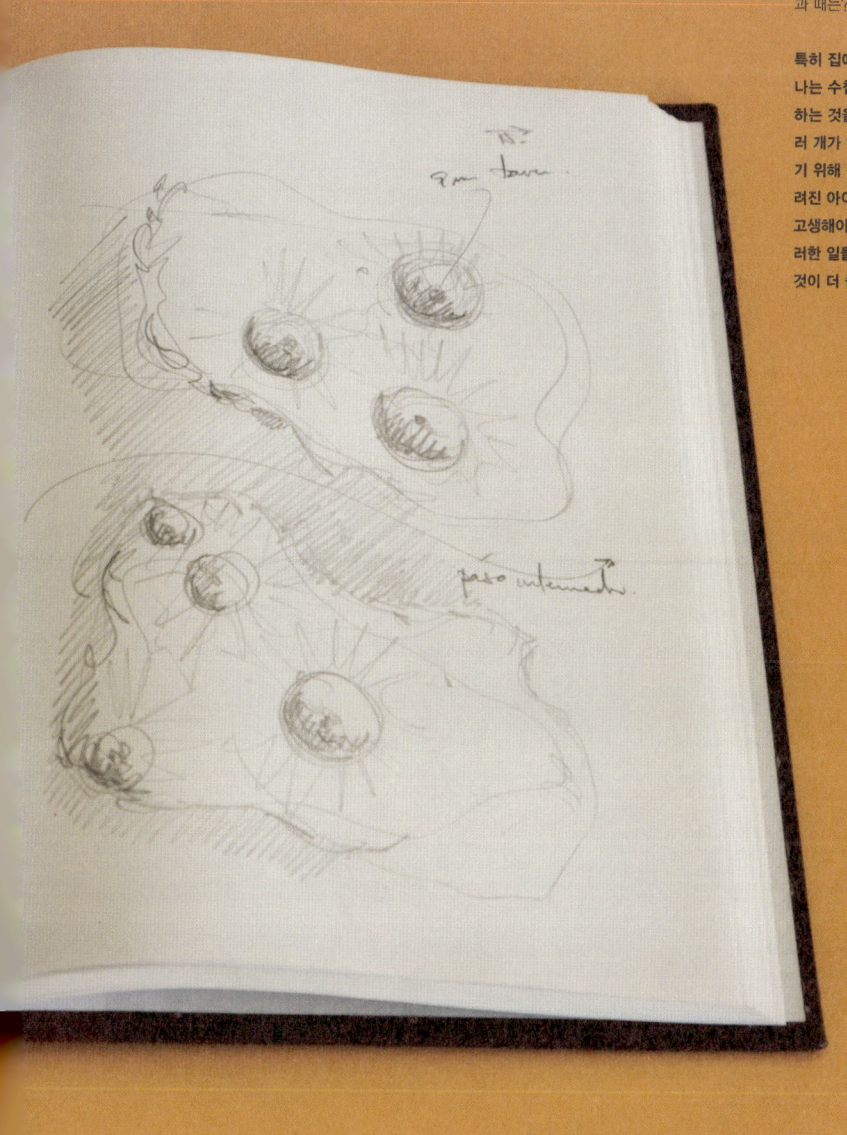

수첩을 가장 많이 사용하는 공간과 때는?

특히 집에서 밤에 많이 사용한다. 나는 수첩을 하나만 가지고 사용하는 것을 잘 못해서 집 곳곳에 여러 개가 있다. 지난 아이디어를 찾기 위해 각각의 스케치북마다 그려진 아이디어 조각들을 찾기 위해 고생해야 할 때도 있다. 하지만 이러한 일들이 스케치북을 다시 보는 것이 더 즐거워지게 만들어준다.

What influence does a notebook have in your projects and life as an architect?

It gives me freedom to think. When you start wit other process in the design (computer, calculations,etc..) everything seems to be controlled not only by you, but by the technical process.

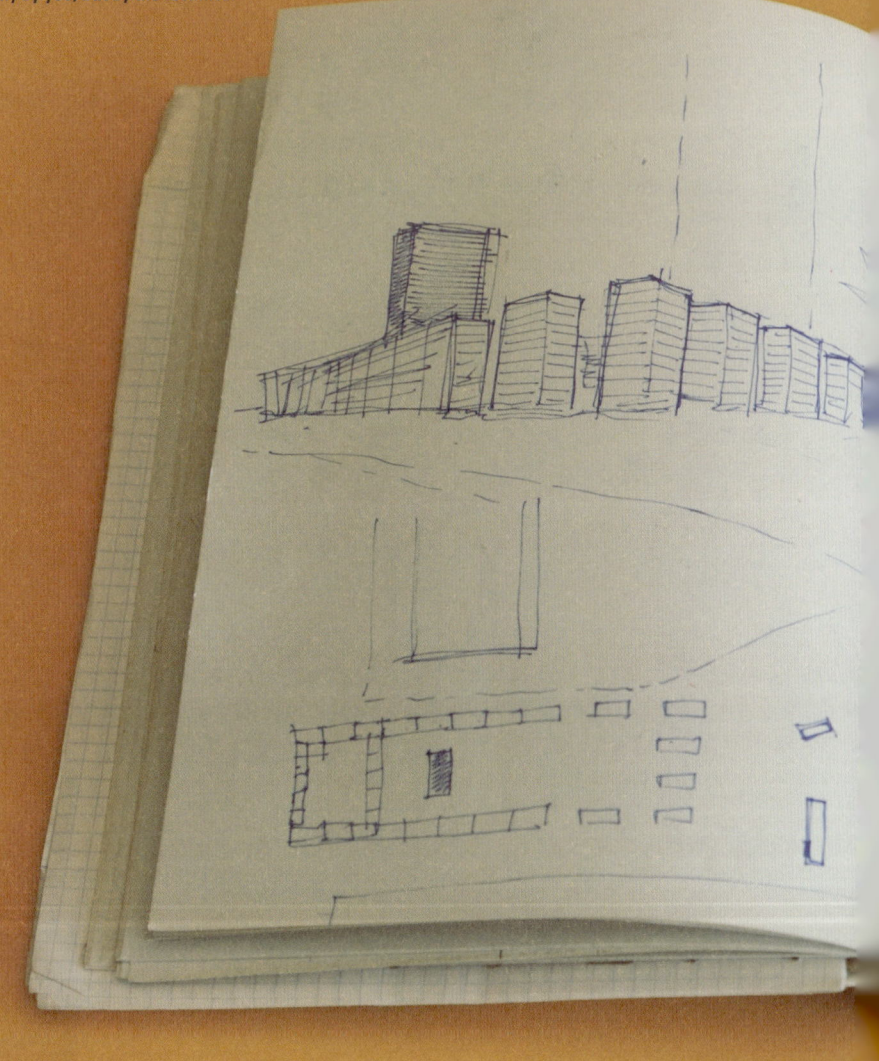

It gives me freedom

수첩은 당신의 작품과 삶에 어떤 영향을 끼치는가?

수첩은 나에게 생각의 자유를 준다. 컴퓨터나 계산을 통해 디자인을 시작할 때에는 나만이 아닌, 기계적인 과정이 지배하는 것 같다.

to think.

Are there anything else other than a notebook that you use to keep a record of your thoughts and ideas?

I always try to sketch at the same time in 2D and 3D, so I usually work with conceptual models. I keep those ideas in 3D around me during the whole process of the project. Apart from that, once the idea of the project is in an advanced process **I love to contact an artist to make his own interpretation of the work.** Those objects created became a part of my daily life.

수첩 외에 자신의 생각을 기록하는 방법과 도구는 무엇이 있는가?

나는 항상 2D와 3D를 함께 생각하고 그리기위해 노력하기 때문에 컨셉 모형들로 작업을 많이 하는 편이다. 프로젝트를 작업하는 내내 3D로 만들어진 아이디어들을 내 주변에 둔다. 모형 외에도, **프로젝트가 좀 더 발전 했을 때에 예술가를 불러 그 사람이 재해석한 나의 프로젝트를 보는 것을 좋아한다.** 이러한 과정에서 만들어진 것들은 나의 일상속의 일부분이 된다.

A battle is raging even if you're not aware of it

"A battle is raging even if I'm not aware of it. It is raging wether I like it or not.
I want to hide, disappear, run away from here. I want to return to my shelter but I can not, they will not let me. Do not know them, not the face they have, only know they are using weapons against words, threats against arguments, fears against dreams, fear is their great weapon and I can not feel safe anymore.
What if in the end I'm able to hide? who will defend me? Perhaps the ignorance dressed of arrogance that governs us. No, we were never a priority to them.
A battle is raging even I don't understand, is raging and I'm the enemy.
No, I will not hide, protect only me would be like to becoming a faceless man. I'll keep using words, arguments and dreams to try to build a shelter, a shelter made by all and for all. A refuge where not necessary to hide, where fear can not catch us. A shelter made to resist the storm when it returns, because it surely will return.
a battle is raging and I am aware, is raging and I will win."

-Miguel Arraiz García

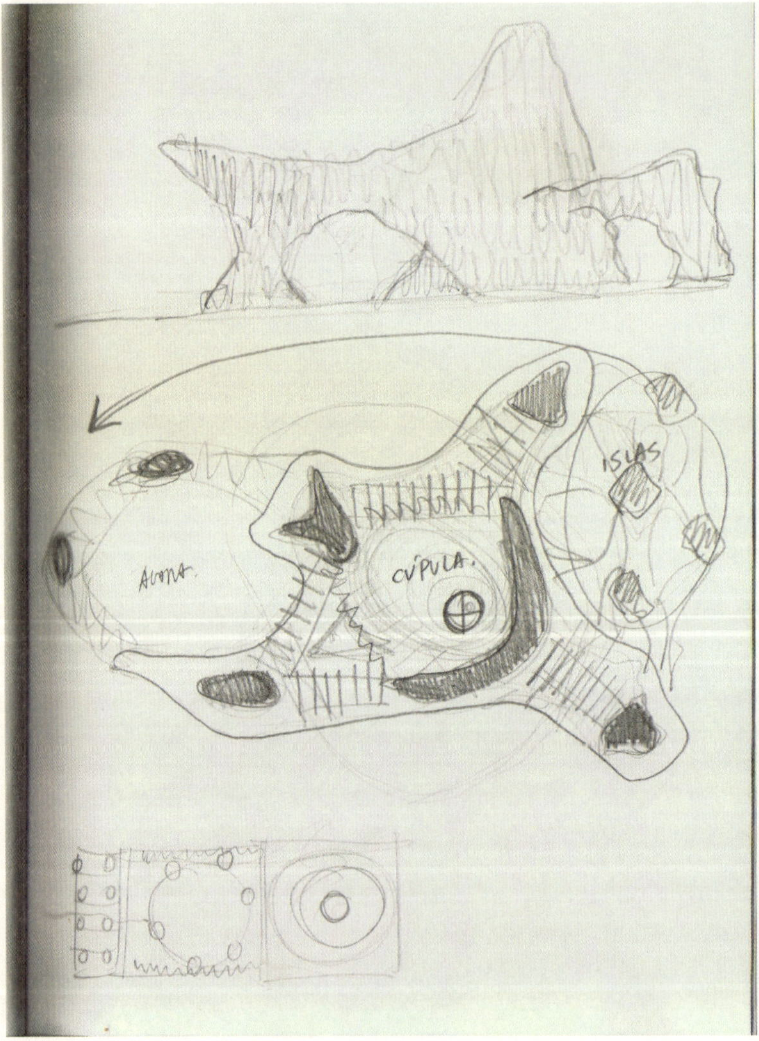

Don't pay attention to the man behind the curtain

"We live surrounded by images, ideas trying to get our attention, in which superficial prepends content and artificiality prepends reality. We opt for certain icons for purely aesthetic, often unaware that behind this great show we attend there is ideology. Life is about choices, and what data do we have to take them? Do we look beyond aesthetics? Do we believe in the shadows and lights that produces the show ore we left the cave and find out who is behind them and who produces them?

We worship the golden calf, "aware" that it is not real, but rather we prefer to live a comfortable lie. No care about the search or the way we should follow, we want pleasure and we want it now and now. We revere and fear the unknown, unable to grasp its reality, we actually prefer giving it "divine" features and transfer the lack of faith in ourselves in an excessive, irrational faith in certain icons.

Or we follow the golden path, without losing the ability to look outside of it, where real life is, where important things are. And at the end of the road be able to deal with the false myths created, check out who is really behind the aesthetics though often we will discover that behind there are no ethics, even truth hurts.

Although once discovered the true Wizard of Oz continues shouting "Pay no attention to that man behind the curtain," we will be as Dorothy and we will continue to seek our truth, the hidden one behind the curtains."

-Miguel Arraiz García

www.noelarraiz.com

www.noelarraiz.com

www.noelarraiz.com

I have nostalgia of future

"I gave up, I have to admit, I gave no more. My young rebellious spirit gave way to hackneyed phrases. That if any past was better, that if life is three days, if you were my age would understand . . . and the worst was not giving up but think that those still fighting were wrong.
I saw these young people struggle with hope and a certain innocence, while recognizing that innocence was becoming less or at least more fleeting. They were idealistic, had everything to do, but I saw what the future held for them, looked beyond the bubble in which we tried to protect them. I watched as others took advantage of all that, accumulated, devastated, destroyed, deflated illusions. Getting them to surrender to use their own hands to help them continue accumulating, devastating and destroying. The future could not be better, we were doomed to be the pawns of the great mound, whose queen was always hidden
But I woke up and saw that the past was only a possible future, and there were many possible futures, futures that were in our hands if we don't surrender. I saw that together we could build without destroying that who governs us can not remain hidden, that bubbles are of illusion and can grow to infinity. And I had nostalgia, I had nostalgia for the Future."

-Miguel Arraiz Garcia

Artist marcelo fuentes interpretation of the project

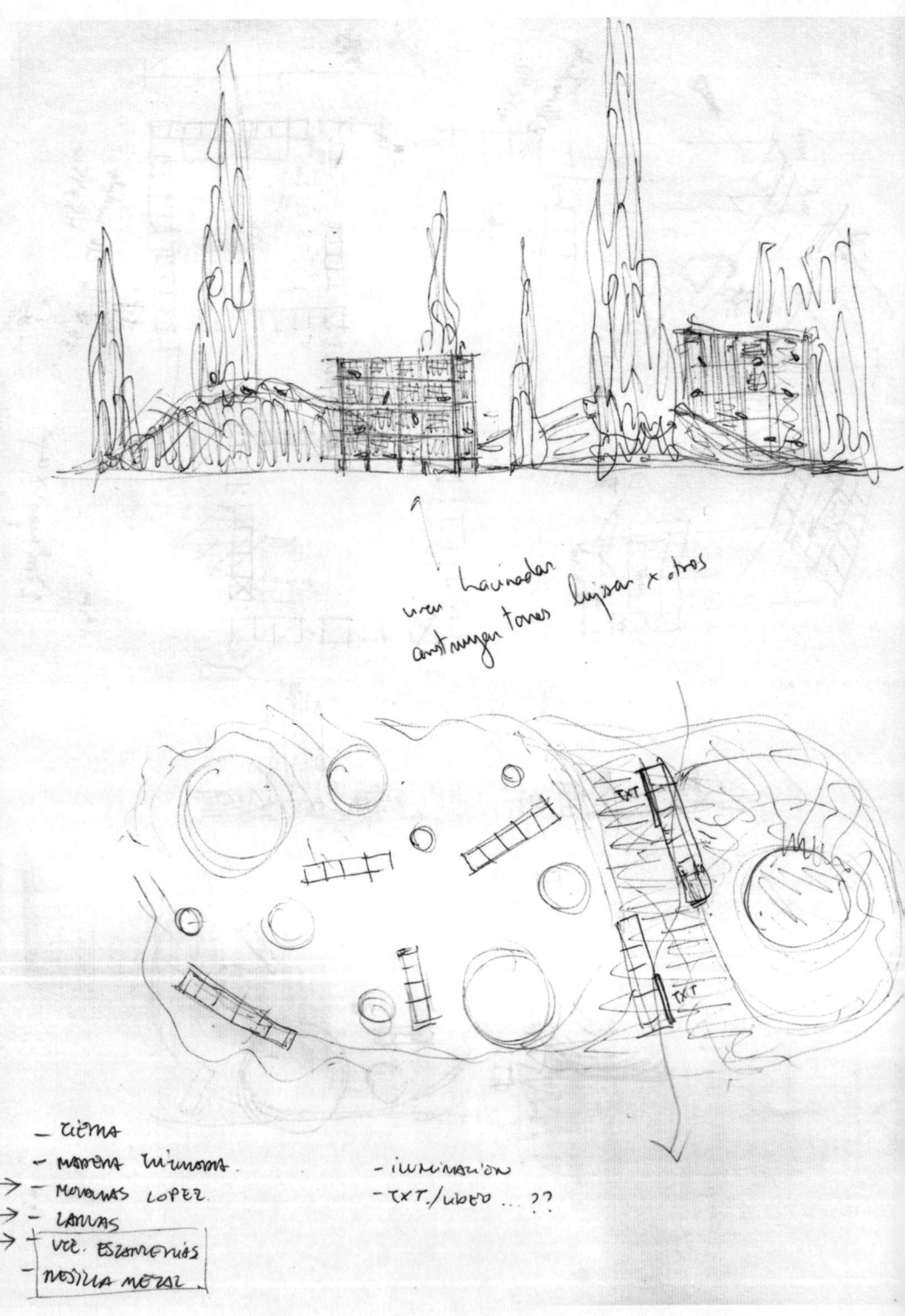

403

INDEX

AA & U
www.urban-a-where.com
220

ARHITEKTURA d.o.o.
www.arhitektura-doo.si
110

ARPHENOTYPE
www.arphenotype.com
www.horhizon.com
www.digitales-gestalten.de
194

b4 architects
www.b4architects.com
156

BOARD
www.b-o-a-r-d.nl
148

CHA:COL
www.chacol.net
50

DIMOS MOYSIADIS
dmoysiadis.gr
232

Donner Sorcinelli Architecture
www.donner-sorcinelli.it
98

eu.k Architects
www.eukarchitects.com
132

EXTERNAL REFERENCE ARCHITECTS
www.externalreference.com
242

Gambardellarchitetti
www.gambardellarchitetti.com
88

Gutiérrez-delaFuente Arquitectos
www.gutierrez-delafuente.com
26

Heather Woofter [Axi:Ome]
www.axi-ome.net
182

Hybrid Space Lab
www.hybridspacelab.net
124

IAD PARTNERS
www.groupiad.com
370

IaN+
www.ianplus.it
316

Jongyeon Bahk [Grid-A]
architour.pe.kr / grid-a.net
278

LIMA URBAN LAB
www.lul-lab.com
344

Miguel Arraiz García [Bipolaire Arquitectos / Pink Intruder]
www.bipolaire.net
www.pinkintruder.com
384

object-e architecture
object-e.net
300

OOIIO Architecture
www.ooiio.com
260

OSA
www.o-s-a.com
206

XARIS TSITSIKAS
x-t.gr
232

YOSHIHARA McKEE ARCHITECTS
www.yoshiharamckee.com
70

ZO_LOFT ARCHITECTURE & DESIGN
www.zo-loft.com
356